Numerical Methods for Unsteady Compressible Flow Problems

Numerical Analysis and Scientific Computing Series

Series Editors:
Frederic Magoules, Choi-Hong Lai

About the Series
This series, comprising of a diverse collection of textbooks, references, and handbooks, brings together a wide range of topics across numerical analysis and scientific computing. The books contained in this series will appeal to an academic audience, both in mathematics and computer science, and naturally find applications in engineering and the physical sciences.

Iterative Splitting Methods for Differential Equations
Juergen Geiser

Handbook of Sinc Numerical Methods
Frank Stenger

Computational Methods for Numerical Analysis with R
James P Howard, II

Numerical Techniques for Direct and Large-Eddy Simulations
Xi Jiang, Choi-Hong Lai

Decomposition Methods for Differential Equations
Theory and Applications
Juergen Geiser

Mathematical Objects in C++
Computational Tools in A Unified Object-Oriented Approach
Yair Shapira

Computational Fluid Dynamics
Frederic Magoules

Mathematics at the Meridian
The History of Mathematics at Greenwich
Raymond Gerard Flood, Tony Mann, Mary Croarken

Modelling with Ordinary Differential Equations: A Comprehensive Approach
Alfio Borzì

Numerical Methods for Unsteady Compressible Flow Problems
Philipp Birken

For more information about this series please visit: https://www.crcpress.com/ Chapman--HallCRC-Numerical-Analysis-and-Scientific-Computing-Series/book-series/ CHNUANSCCOM

Numerical Methods for Unsteady Compressible Flow Problems

Philipp Birken

CRC Press
Taylor & Francis Group
Boca Raton London New York

CRC Press is an imprint of the
Taylor & Francis Group, an **informa** business
A CHAPMAN & HALL BOOK

by CRC Press
6000 Broken Sound Parkway NW, Suite 300, Boca Raton, FL 33487-2742

and by CRC Press
2 Park Square, Milton Park, Abingdon, Oxon, OX14 4RN

Library of Congress Cataloging-in-Publication Data

Names: Birken, Philipp, author.
Title: Numerical methods for unsteady compressible flow problems / Philipp Birken.
Description: First editon. | Boca Raton : C&H\CRC Press, 2021. | Series: Chapman & Hall/CRC numerical analysis and scientific computing series | Includes bibliographical references and index.
Identifiers: LCCN 2021000979 (print) | LCCN 2021000980 (ebook) | ISBN 9780367457754 (hardback) | ISBN 9781003025214 (ebook)
Subjects: LCSH: Viscous flow--Mathematical models. | Unsteady flow (Fluid dynamics)--Mathematical models. | Navier-Stokes equations--Numerical solutions. | Computational fluid dynamics.
Classification: LCC QA929 .B57 2021 (print) | LCC QA929 (ebook) | DDC 532/.58--dc23
LC record available at https://lccn.loc.gov/2021000979
LC ebook record available at https://lccn.loc.gov/2021000980

ISBN: 978-0-367-45775-4 (hbk)
ISBN: 978-1-032-02183-6 (pbk)
ISBN: 978-1-003-02521-4 (ebk)

Typeset in CMR10
by KnowledgeWorks Global Ltd.

Für Margit

Contents

Preface and acknowledgments

This book was written mostly in Kassel, and while traveling. In planes, during conferences (I remember a particularly productive session on Hawaii after the AIAA conference), on vacation in France, and a lot while sitting in the trains of the Deutsche Bahn. The ICE high speed train provides excellent working conditions, I have to say, and in this I include the bad internet connection. Sometimes it helps to have less distractions. It was finished in Lund during the COVID-19 pandemic. A strange time.

There can be many motivations to write a book. I met Pieter Wesseling only once as a PhD student at a conference. He said that he wrote his books to give them to his PhD students, as a basis for their work. In a treatise on academic vanity, the author wrote tongue in cheek, that he wrote it to see his name once more in print. Peter S. Beagle, the author of the last unicorn, denied most requests for writing book blurb using the argument that if he wanted to read a book, he would write one. In a way, this applies even more to nonfiction. Writing allows to organize the material in ones own mind as few other processes, if at all.

In the field of computational fluid dynamics (CFD), continuous improvements in algorithms and hardware have brought us to a situation where unsteady flow computations are perfectly feasible and accurate computations of three-dimensional turbulent flows around complex geometries are within reach. On the hardware side, this is enabled by massive parallelism. Regarding algorithms, important progress has been made on high-order discretizations such as Discontinuous Galerkin, making Large Eddy Simulations (LES) feasible. Furthermore, the last 20 years have given tremendous improvement of iterative solvers such as multigrid for the arising huge nonlinear systems within implicit time integration, specifically for the compressible Navier-Stokes equations.

There are many very good books on CFD, on turbulence, on finite volume methods, on high order methods, on methods for hyperbolic problems, for viscous problems, and so on. There are also many very good books on numerical methods for time integration and iterative solvers. But none of them combines modeling, high-order discretizations in space and time and state-of-the-art iterative solvers for the compressible Navier-Stokes equations. This being the book that I wanted to read, I wrote it. I hope that other people find it useful as well.

It is written with both mathematicians and engineers in mind. The idea of the text is to explain the important ideas to the reader, while giving enough detail and pointers to literature to facilitate implementation of methods and application of concepts for analysis and construction of numerical methods. Hereby, I tried to reconcile the language used in the two communities, in particular when it comes to multigrid methods, where independent method development went on for decades.

A project like this would not have been possible without help from many people. First of all, I'd like to mention the German Research Foundation (DFG), which funded my research during 2006–2012 as part of the collaborative research area SFB/TR TRR 30. This interdisciplinary project has provided invaluable perspectives. It also allowed me to hire students to help me in my research and I'd like to thank Benjamin Badel, Maria Bauer, Veronika Diba, Marouen Ben Said, and Malin Trost, who did a lot of coding and testing and created a lot of figures.

I'd like to thank all my collaborators over the past years. Collaboration is fun and it makes us do better research. You are all part of this book. In particular, thanks to my former PhD advisor Andreas Meister. A heartfelt cheerio to Antony Jameson, from whom I have learned so much about CFD and about being a researcher. Not to forget Gregor Gassner, who had important influence on the parts about high order methods. Furthermore, thanks to Mark H. Carpenter and Gustaf Söderlind for a lot of interesting discussions, which have greatly impacted my thinking.

I'd also like to thank my colleagues in Kassel and Lund for a friendly atmosphere. Furthermore, there are the colleagues at Stanford at the Institute of Mathematical and Computational Engineering and the Aero-Astro department, where I always encounter a friendly and inspiring atmosphere.

And not least, Sigrun Ortleb, Viktor Linders and my old friend Tarvo Thamm, for proofreading.

Finally, my Margit who makes everything worthwhile.

Lund, December 20, 2020

Chapter 1

Introduction

Historically, the computation of steady flows has been at the forefront of the development of computational fluid dynamics (CFD). It began with the design of rockets and the computation of the bow shock at supersonic speeds and continued with the aerodynamic design of airplanes at transonic cruising speed [143]. Only since the 2000s, increasing focus has been put on unsteady flows, which are more difficult to compute. This has several reasons. First, computing power has increased dramatically, and for 5,000 euros it is now possible to obtain a machine that is able to compute about a minute of real-time simulation of a nontrivial unsteady three-dimensional flow in a day. As a consequence, ever more nonmilitary companies are able to employ numerical simulations as a standard tool for product development, opening up a large number of additional applications. Examples are the simulation of tunnel fires [24], flow around wind turbines [304], fluid-structure-interaction like flutter [89], aeroacoustics [63], turbomachinery, flows inside nuclear reactors [233], wildfires [232], hurricanes and unsteady weather phenomena [231], gas quenching [178] and many others. More computing capacities will open up further possibilities in the next decade, which suggests that the improvement of numerical methods for unsteady flows should start in earnest now. Finally, the existing methods for the computation of steady states, while certainly not at the end of their development, have matured, making the consideration of unsteady flows interesting for a larger group of scientists. This research monograph is written with both mathematicians and engineers in mind, to give them an overview of the state of the art in the field and to discuss a number of new developments, where we will focus on the computation of compressible viscous flows, as modeled by the compressible Navier-Stokes equations.

These form the most important model in fluid dynamics, from which a number of other widely used models can be derived, for example the incompressible Navier-Stokes equations, the Euler equations or the shallow water equations. An important feature of fluids that is present in the Navier-Stokes equations is turbulence, which roughly speaking appears if the Reynolds number of the problem at hand is large enough. Since Reynolds numbers for air flows as modeled by the compressible equations are large, most airflows of practical importance are turbulent.

Another important feature of the Navier-Stokes equations is the boundary layer, which makes it necessary to use very fine grids. Since explicit time integration methods have an inherent stability constraint, they need to choose their time step on these grids based on stability only and not via error control. This makes the use of implicit time integration methods desirable, since these are not bound by stability constraints as explicit schemes. However, using implicit schemes requires solving linear or nonlinear equation systems. Consequently, a large part of this book will be about implicit schemes and schemes for solving linear and nonlinear equation systems.

1.1   The method of lines

Here, we follow the method of lines paradigm:

1. The space discretization transforms the partial differential equation (PDE) into a system of ordinary differential equations (ODE), introducing a discretization error.

2. The implicit time integration transforms the ODE into a system of algebraic equations, introducing a time integration error, that needs to be related to the discretization error.

3. If this algebraic system is nonlinear, it is approximately solved by an iterative method (until termination), which introduces an iteration error that needs to be smaller than the time discretization error.

4. If this algebraic system is linear or if Newton's method was chosen in the previous step, another iterative method solves the linear system (until termination), which introduces another iteration error, which needs to be related to the time discretization error, respectively the error in the nonlinear solver.

The main advantage of the method of lines is its immense flexibility, which allows the reuse of existing methods that can be tested and analyzed for the simpler subproblems, as well as the simple exchange of any part of the overall solution procedure. Therefore, its use is ubiquitous in scientific computing, which means that most of the techniques discussed here can be applied to a large class of problems, as long as they are modeled by time-dependent partial differential equations. On the other hand, the main drawback of the approach is in the limitation of how spatial and temporal adaptation can be connected. Therefore, space-time discretizations are an interesting alternative.

Thus, we arrive at the common questions of how we can guarantee robustness and accuracy, obtain efficient methods and all of that at reasonable implementation cost.

Regarding space discretization, the standard methods for flow problems are finite volume (FV) methods. These respect a basic property of the Navier-Stokes equations, namely conservation of mass, momentum and energy, and despite significant gaps in theory for multidimensional problems, it is known how to obtain robust schemes of order up to two. Now, for high-resolution turbulent flow computations, these orders are not enough. The question of obtaining higher-order schemes is, however, rather challenging and has been an extremely active area of research for the past 20 years. As the most prominent type of methods there, discontinuous Galerkin (DG) schemes have been established. These schemes use higher-order polynomials and a Galerkin condition to determine the solution. However, as for all schemes of higher order, the question of efficiency for the Navier-Stokes equations is still the topic of much ongoing research, in particular for implicit time integration.

In the case of an explicit time integration, explicit Runge-Kutta methods are the best choice. In the implicit case, BDF methods are used widely in industry, essentially because these just need to solve one nonlinear system per time step. However, third- and higher-order BDF schemes have limited stability properties. Furthermore, time adaptivity is a problematic issue. This is important for an implicit scheme, since the whole point is to choose the time step based on error estimators and not on stability constraints. Therefore, singly diagonally implicit Runge-Kutta (SDIRK) schemes are the important alternative, which was shown to be competitive with BDF methods at engineering tolerances [23]. These consist of a sequence of backward Euler steps with different right-hand sides and time step sizes, thus allowing a modular implementation. Furthermore, time adaptivity can be done at almost no additional cost using embedded schemes. Another interesting alternative, in particular for more accurate computations, includes Rosenbrock schemes that in essence are linearized SDIRK methods, thus only needing linear system solves.

The steps that mainly determine the efficiency of an implicit time integration scheme are the last two in the method of lines: Solving equation systems. In the industry but also in academia, code development has been driven by the desire to compute steady flows. This requires the solution of one large algebraic system and the fastest codes to do so use a multigrid method. The use of multigrid was pioneered in CFD by Antony Jameson, and the current state of the art is in large parts due to his further development of the method. In fact, the majority of codes used in industry employ this strategy. The multigrid method for steady states can be carried over to unsteady flows with very little additional coding effort. This makes in these codes steady-state multigrid the method of choice even for unsteady flows. For finite volume discretizations, the situation is that for the Euler equations, convergence rates as low as 0.1 have been achieved for flows around airfoils. During the last ten years important progress has been made on RANS equations as well, which has given convergence rates of about 0.8, when before that it was closer to 0.99. With regards to DG discretizations, the design of good multigrid methods is work in progress.

The main alternative is Newton's method, which requires solving sequences of linear systems. Therefore, common data structures needed are vectors and matrices. Since an explicit code is typically based on cell or point-based data structures and not on vectors, the implementation cost of this type of methods is considered to be high. Together with the fact that the canonical initial guess for the steady state (freestream data) is typically outside the region of convergence of Newton's method, this has led to a bad reputation of the method in the CFD community.

Now, if we consider a steady-state equation, discretized in space, we obtain a nonlinear algebraic equation of the form

$$\underline{\mathbf{f}}(\underline{\mathbf{u}}) = \underline{\mathbf{0}},$$

with $\underline{\mathbf{u}} \in \mathbb{R}^m$ and $\underline{\mathbf{f}} : \mathbb{R}^m \to \mathbb{R}^m$. However, the nonlinear equation arising from an implicit time discretization in the method of lines is of the form

$$(\underline{\mathbf{u}} - \underline{\mathbf{s}})/(\alpha \Delta t) - \underline{\mathbf{f}}(\underline{\mathbf{u}}) = \underline{\mathbf{0}},$$

where $\underline{\mathbf{s}}$ is a given vector, α a method-dependent parameter, and Δt is the time step size. Due to the additional term, we expect the second system to be easier to solve, although this advantage diminishes with increasing time step size.

When considering standard multigrid codes, it turns out that, if at all, the convergence speed is increased only slightly when going from steady to unsteady flows. For Newton's method in an implicit time integration, the performance improves dramatically. The reason is that the canonical initial guess is the solution from the last time level. This is in a sense close to the solution at the new time level, which allows us to make use of the quadratic convergence right away.

Furthermore, when solving the linear equation systems using a Krylov subspace method like GMRES, the Jacobian is needed in matrix vector products only. Since the matrix is a Jacobian, it is possible to replace these by a finite difference approximation. Thus, a method is obtained that does not need a Jacobian and needs no additional implementation effort when changing the spatial discretization, just in the spirit of the flexibility of the method of lines. Unfortunately, this is not completely true in that Krylov subspace methods need a preconditioner to be truly efficient. It is here that a lot of research has been put in and more research needs to be done to obtain efficient, robust, and easy-to-implement schemes. Summarizing, it is necessary to reevaluate the existing methods for unsteady flows.

As for preconditioning, the linear systems that arise have unfortunately few properties that can be exploited; in particular, they are nonnormal and not diagonally dominant for reasonable Δt. However, the systems have a block structure arising from grouping unknowns from the same cell together. For finite volume schemes, where the blocks are of size four for two-dimensional flows and size five for three-dimensional flows, a block incomplete LU (ILU) decomposition leads to a fast method. This situation is different for DG schemes. There, the number of unknowns in one cell depends on the dimension, the order of the scheme, and the particular method chosen. As a rule of thumb, the number of unknowns in two- and three-dimensional flows is in the dozens and hundreds, respectively. This makes a significant difference for efficiency with the result being that currently, the class of problems where an implicit scheme is more efficient than an explicit one is significantly smaller for DG schemes than for FV schemes. For a specific DG scheme, namely a modal method with a hierarchical polynomial basis, we describe a preconditioner that is able to address this specific problem.

Repeating the issues named earlier about robustness, efficiency, and accuracy, an important topic arises: Since an implicit method requires one or two subiterations, there are a lot of parameters to choose in the method. Here, the goal is to have only one user-defined parameter in the end, namely the error tolerance, and all other parameters should be determined from there. Furthermore, the termination criteria in the subsolver should be chosen such that there is no interference with the discretization error, but also no oversolving, which means that these schemes should not perform more iterations than necessary. Finally, iterative methods have the inherent danger of divergence. Therefore, it is necessary to have feedback loops on what happens in these cases. Otherwise, the code will not be robust and, therefore, not be used in practice. In these regards, mathematics is a tremendous help, since it allows to obtain schemes that are provably convergent, to derive termination criteria that are sharp and to detect divergence.

1.2 Hardware

A fundamental trend in computing since the introduction of the microprocessor has been Moore's law, which means that the speed of CPUs increases exponentially. There is a second important trend, namely that instead of one computation on a single processor, multiple computations are carried out in parallel. This is true for huge clusters commonly used in high-performance computing, for common PCs that have multiple CPUs with multiple cores and for graphics processing units (GPUs) that are able to handle a thousand threads in parallel. This means that numerical methods must be devised to perform well in parallel. Typically, this trend was driven by hardware manufacturers not having numerical computations in mind; then compilers and operating systems were built by computer scientists not having numerical computations in mind. This leads to a situation where the numerical analysts at the bottom of the food chain have a hard time designing optimal methods, in particular when hardware architecture is quickly changing as it is now.

However, there is an interesting development going on with GPUs, which were originally designed for efficient single precision number crunching. When the market leader Nvidia realized that scientific computation is an attractive market, it developed GPUs able of performing double precision computations together with making coding on GPUs easier using the documentation of CUDA. This makes a programming paradigm, where codes are parallelized on multicore CPUs using the MPI standard and on top of that, GPUs are employed.

Another important trend is that while the available memory has steadily increased as well over the years, this has happened at a slower rate than the increase in speed of CPUs. As a result, memory per core has been decreasing. Summarizing, independently of the specific future hardware, it will feature more and more parallelism with less and less memory per core. This has to be reflected in the design of numerical methods, in particular for simulations of three-dimensional turbulent flows.

1.3   Notation

Throughout this book, we will use bold capital letters ($\mathbf{A}$) to indicate matrices and bold lowercase letters ($\mathbf{x}$) to indicate vectors. Lowercase letters represent scalars, and thus the components of vectors are lowercase letters with indices. Thus, $\mathbf{u}_1$ is the first vector of a family of vectors, but u_1 would be the first component of the vector $\mathbf{u}$. Specifically, the three space directions are x_1, x_2, and x_3 as the components of the vector $\mathbf{x}$. A vector with an underbar $\underline{\mathbf{u}}$ denotes a vector representing the discrete solution on the complete grid.

1.4   Outline

This book is an updated and corrected version of my habilitation thesis [25]. The content essentially follows the order of steps in the method of lines. First, we describe in chapter 2 the basic mathematical models of fluid dynamics and discuss some of their properties. Then we discuss space discretization techniques in chapter 3, in particular finite volume and DG methods. In chapter 4, different time integration techniques are presented, in particular explicit and implicit schemes, but also schemes that do not fall directly in the method of lines context. If implicit schemes are used, this results in linear or nonlinear equation systems; and the solution of these is discussed in chapter 5. In particular, fixed-point methods, multigrid methods, Newton methods, and Krylov subspace methods as well as Jacobian-Free Newton-Krylov methods are described. Preconditioning techniques for Krylov subspace methods are described in chapter 6. In chapter 7, we summarize and combine the techniques of the previous chapters and describe how the final flow solvers look like, as well give an assessment of their performance. Finally, unsteady thermal Fluid-Structure interaction is considered in chapter 8, and the application of the techniques discussed before is described.

Chapter 2

The governing equations

We will now describe the equations that will be used throughout this book. The mathematical models employed in fluid mechanics have their basis in continuum mechanics, which means that it is not molecules that are described, but a large number of those molecules that act as if they were a continuum. Thus velocities, pressure, density, and similar quantities are of a statistical nature and say that on average, at a certain time, the molecules in a tiny volume will behave in a certain way. The mathematical model derived through this approach are the Navier-Stokes equations. The main component of these is the momentum equation, which was found in the beginning of the 19th century independently of each other by Navier [211], Stokes [261], Poisson [225], and de St. Venant [69]. During the course of the century, the equations were given more theoretical and experimental backing, so that by now, the momentum equation together with the continuity equation and the energy equation are established as the model describing fluid flow.

This derivation also shows one limitation of the mathematical model, namely for rarefied gases as in outer layers of the atmosphere, the number of molecules in a small volume is no longer large enough to allow statistics.

From the Navier-Stokes equations, a number of important simplifications have been derived, in particular the incompressible Navier-Stokes equations, the Euler equations, the shallow water equations or the potential flow equations. We will discuss in particular the Euler equations, which form an important basis for the development of discretization schemes for the Navier-Stokes equations, as well as an already very useful mathematical model in itself.

2.1 The Navier-Stokes equations

The Navier-Stokes equations describe the behavior of a Newtonian fluid. In particular, they describe turbulence, boundary layers, as well as shock waves and other wave phenomenas. They consist of the conservation laws of mass, momentum and energy. Thus they are derived from integral quantities, but for the purpose of analysis, they are often written in a differential form. A more detailed description can be found for example in the textbooks of Chorin and Marsden [59] and Hirsch [135].

In the following sections, we will start with dimensional quantities, denoted by the superscript ̂, and derive a nondimensional form later. We now consider an open domain $\mathcal{U} \subset \mathbb{R}^d$, and the elements of $\mathcal{U}$ are written as $\mathbf{x} = (x_1, ..., x_d)^T$.

2.1.1   Basic form of conservation laws

The conservation of a quantity is typically described by rewriting the amount $\phi_\Omega(t)$ given in a control volume $\Omega \subset \mathcal{U}$ using a local concentration $\psi(\mathbf{x}, t)$:

$$\phi_\Omega(t) = \int_\Omega \psi(\mathbf{x}, t) d\Omega.$$

Conservation means that the amount ϕ_Ω can only be changed by transport over the boundary of Ω or internal processes. An important tool to describe this change is Reynolds' transport theorem.

Theorem 1 (Reynolds' transport theorem) *Let $\Omega(t)$ be a possibly time-dependent control volume, ψ a differentiable function and $\mathbf{v}(\mathbf{x}, t)$ be the velocity of the flow. Then*

$$\frac{d}{dt} \int_\Omega \psi d\Omega = \int_\Omega \partial_t \psi d\Omega + \int_{\partial\Omega} \psi \mathbf{v} \cdot \mathbf{n} ds. \tag{2.1}$$

The proof is straightforward using multidimensional analysis; see e.g. [302, page 10].

The view of the fluid taken here is that the observer is fixed in space, whereas the fluid is not, similar to sitting on the bank of a river and watching it flow by. This is called an Eulerian approach, as opposed to a Lagrangian one. In the latter, which is commonly used in solid mechanics, we explicitly track the path of specific particles. Here, we instead keep track of the properties of the fluid in a given domain by accounting for what flows into the domain and what flows out.

We will now consider control volumes that are fixed in time. Thus we have

$$\int_\Omega \partial_t \psi d\Omega = \partial_t \int_\Omega \psi d\Omega.$$

Furthermore, we need the famous Gaussian divergence theorem.

Theorem 2 (Divergence theorem) *For any volume $\Omega \in \mathbb{R}^d$ with piecewise smooth closed boundary $\partial\Omega$ and any differentiable vector field $\mathbf{u}$, we have*

$$\int_\Omega \nabla \cdot \mathbf{u} d\Omega = \int_{\partial\Omega} \mathbf{u} \cdot \mathbf{n} ds.$$

To formulate a conservation law, we start with the right-hand side in (2.1) and then need to specify what this term is equal to. Besides the flow over the boundary already present there, there can be sinks, sources, and forces. In the absence of sinks and sources, we thus need to specify surface and volume forces acting on the fluid.

2.1.2 Conservation of mass

The concentration of mass is its density $\hat{\rho}$. Since forces do not change mass, the *conservation equation of mass* (also called continuity equation) can be written for an arbitrary control volume Ω as

$$\partial_{\hat{t}} \int_{\Omega} \hat{\rho} d\Omega + \int_{\partial\Omega} \hat{\mathbf{m}} \cdot \mathbf{n} ds = 0. \tag{2.2}$$

Here $\hat{\mathbf{m}} = \hat{\rho}\mathbf{v}$ denotes the momentum vector divided by the unit volume. Since this is valid for any Ω, application of the Gaussian theorem for the boundary integral leads to the differential form

$$\partial_{\hat{t}}\hat{\rho} + \nabla \cdot \hat{\mathbf{m}} = 0. \tag{2.3}$$

Here, the divergence operator is meant with respect to the spatial variables only, throughout the book.

Note that by contrast to the two following conservation laws, which are based on fundamental principles of theoretical physics like the Noether theorem, conservation of mass is not a proper law of physics, since mass can be destroyed and created. A particular example is radioactive decay, where mass is transformed into energy, meaning that the underlying physical law here is conservation of energy via Einstein's discovery of the equivalence of mass and energy. However, for the purpose of nonradioactive fluids at nonrelativistic speeds, (2.2) is a perfectly reasonable mathematical model.

2.1.3 Conservation of momentum

The equation for the *conservation of momentum* is based on Newton's second law, which states that the change of momentum in time is equal to the acting force. For the time being, we assume that there are no volume or external forces acting on the fluid and look at surface forces only. As relevant terms we then have the surface pressure, which results in a force, and the forces resulting from shear stresses due to viscous effects. Additionally, we get the flow of momentum from Reynolds' transport theorem. With the pressure $\hat{p}$, we obtain for an arbitrary control volume Ω

$$\partial_{\hat{t}} \int_{\Omega} \hat{m}_i d\Omega + \int_{\partial\Omega} \hat{m}_i \hat{\mathbf{v}} \cdot \mathbf{n} ds = \int_{\partial\Omega} (\hat{\mathbf{s}}_i - \hat{p}\mathbf{e}_i) \cdot \mathbf{n} ds, \qquad i = 1, ..., d, \tag{2.4}$$

where $\mathbf{e}_i$ is the i unit vector. Again, application of the Gaussian theorem for the boundary integral leads to the differential form

$$\partial_{\hat{t}}\hat{m}_i + \sum_{j=1}^{d} \partial_{\hat{x}_j}(\hat{m}_i\hat{v}_j + \hat{p}\delta_{ij}) = \sum_{j=1}^{d} \partial_{\hat{x}_j}\hat{S}_{ij}, \qquad i = 1, ..., d, \tag{2.5}$$

where δ_{ij} is the Kronecker symbol and the viscous shear stress tensor $\mathbf{S}$ is given by

$$\hat{S}_{ij} = \hat{\mu}\left((\partial_{\hat{x}_j}\hat{v}_i + \partial_{\hat{x}_i}\hat{v}_j) - \frac{2}{3}\delta_{ij}\nabla \cdot \hat{\mathbf{v}}\right), \qquad i,j = 1,...,d, \qquad (2.6)$$

where $\hat{\mu}$ is the dynamic viscosity. The columns of the stress tensor are denoted by $\hat{\mathbf{s}}_i$. In particular, this means that the shear stresses with $i \neq j$ are proportional to the velocity gradient. If this is not the case for a fluid, it is called non-Newtonian. Examples of this are fluids where the viscosity itself depends on the temperature or the velocity, namely blood, glycerin, oil or a large number of melted composite materials. Note that in (2.6), the experimentally well validated Stokes hypothesis is used, that allows to use only one parameter $\hat{\mu}$ for the description of viscosity.

2.1.4   Conservation of energy

Regarding *conservation of energy*, its mathematical formulation is derived from the first law of thermodynamics for a fluid. The first law states that the change in time in total energy in a control volume Ω is given by sum of the flow of heat and work done by the fluid. The heat flow is given by the sum of the convective flux from the Reynolds' transport theorem (2.1) and the viscous flow due to Fourier's law of heat conduction. Furthermore, the work done by the fluid is due to the forces acting on it. Again, we neglect external forces for now and thus we have the pressure forces and the viscous stresses. With $\hat{E}$ being the total energy per unit mass, we thus obtain for an arbitrary control volume:

$$\partial_{\hat{t}}\int_{\Omega}\hat{\rho}\hat{E}d\Omega + \int_{\partial\Omega}(\hat{H}\hat{\mathbf{m}})\cdot\mathbf{n}ds = \int_{\partial\Omega}\sum_{j=1}^{d}\left(\sum_{i=1}^{d}\hat{S}_{ij}\hat{v}_i - \hat{W}_j\right)\cdot\mathbf{n}ds. \qquad (2.7)$$

As before, application of the Gaussian theorem for the boundary integral leads to the differential form

$$\partial_{\hat{t}}\hat{\rho}\hat{E} + \nabla_{\hat{\mathbf{x}}}\cdot(\hat{H}\hat{\mathbf{m}}) = \sum_{j=1}^{d}\partial_{\hat{x}_j}\left(\sum_{i=1}^{d}\hat{S}_{ij}\hat{v}_i - \hat{W}_j\right). \qquad (2.8)$$

$\hat{H} = \hat{E} + \frac{\hat{p}}{\hat{\rho}}$ denotes the enthalpy and $\hat{W}_j$ describes the flow of heat which, using the thermal conductivity coefficient $\hat{\kappa}$, can be written in terms of the gradient of the temperature $\hat{T}$ as

$$\hat{W}_j = -\hat{\kappa}\partial\hat{T}.$$

The total energy per unit mass $\hat{E}$ is given by the sum of internal energy $\hat{e}$ and kinetic energy as

$$\hat{E} = \hat{e} + \frac{1}{2}|\hat{\mathbf{v}}^2|.$$

2.1.5 Equation of state and Sutherland law

The five differential equations not only depend on the variables $\hat{\rho}$, $\hat{\mathbf{m}}$, and $\hat{\rho}\hat{E}$, but also on a number of other factors. These need to be determined to close the system. First of all, the thermodynamic quantities density, pressure, and temperature are related through the ideal gas law

$$\hat{T} = \frac{\hat{p}}{\hat{\rho}\hat{R}} \qquad (2.9)$$

with the specific gas constant $\hat{R}$.

Furthermore, we need an equation for the pressure $\hat{p}$, which is called the equation of state, since it depends on the particular fluid considered. Typically, it is given as a function $\hat{p}(\hat{\rho}, \hat{e})$. For an ideal gas and fluids similar to one, the adiabatic exponent γ can be used to obtain the simple form

$$\hat{p} = (\gamma - 1)\hat{\rho}\hat{e}, \qquad (2.10)$$

which can be derived using theoretical physics. However, for some fluids, in particular some oils, the equation of state is not given as a function at all, but in the form of discrete measurements only.

Finally, the adiabatic exponent γ and the specific gas constant $\hat{R}$ are related through the specific heat coefficients for constant pressure $\hat{c}_p$, respectively constant volume $\hat{c}_v$, through

$$\gamma = \frac{\hat{c}_p}{\hat{c}_v}$$

and

$$\hat{R} = \hat{c}_p - \hat{c}_v.$$

For an ideal gas, γ is the quotient between the number of degrees of freedom plus two and the number of degrees of freedom. Thus, for a diatomic gas like nitrogen, $\gamma = 7/5$ and therefore, a very good approximation of the value of γ for dry air is 1.4. The specific gas constant is the quotient between the universal gas constant and the molar mass of the specific gas. For dry air, this results in $\hat{R} \approx 287 \text{J/Kg/K}$. Correspondingly, we obtain $\hat{c}_p \approx 1010 \text{J/Kg/K}$ and $\hat{c}_v \approx 723 \text{J/Kg/K}$.

The dynamic viscosity $\hat{\mu}$ is related to the temperature via the Sutherland law

$$\hat{\mu} = \hat{\mu}_{\text{ref}}(\hat{T}/\hat{T}_{\text{ref}})^{3/2}\frac{\hat{T}_{\text{ref}} + \hat{S}}{\hat{T} + \hat{S}}. \qquad (2.11)$$

Hereby, $\hat{S}$ is the Sutherland temperature and $\hat{\mu}_{\text{ref}}$ and $\hat{T}_{\text{ref}}$ are reference values. Suitable values for dry air are $\hat{S} = 110K$, $\hat{\mu}_{\text{ref}} = 18 \cdot 10^{-6} \frac{\text{kg}}{\text{m\,s}}$, and $\hat{T}_{\text{ref}} = 273K$.

2.2   Nondimensionalization

An important topic in the analysis of partial differential equations is the nondimensionalization of the physical quantities. This is done to achieve two things. First, we want all quantities to be $\mathcal{O}(1)$ due to stability reasons and then we want scalability from experiments to real-world problems to numerical simulations. For the Navier-Stokes equations, we will obtain several reference numbers like the Reynolds number and the Prandtl number. These depend on the reference quantities and allow this scaling by specifying how reference values had to be changed to obtain solutions for the same Reynolds and Prandtl number.

A nondimensional form of the equations is obtained by replacing all dimensional quantities with the product of a nondimensional variable with a dimensional reference number:

$$\hat{\phi} = \phi \cdot \hat{\phi}_{\text{ref}}. \tag{2.12}$$

Given reference values for the variables length, velocity, pressure, and density ($\hat{x}_{\text{ref}}$, $\hat{v}_{\text{ref}}$, $\hat{p}_{\text{ref}}$, and $\hat{\rho}_{\text{ref}}$), we can define the reference values for a string of other variables from these:

$$\hat{t}_{\text{ref}} = \frac{\hat{x}_{\text{ref}}}{\hat{v}_{\text{ref}}}, \qquad \hat{E}_{\text{ref}} = \hat{H}_{\text{ref}} = \hat{c}_{\text{ref}}^2 = \frac{\hat{p}_{\text{ref}}}{\hat{\rho}_{\text{ref}}}, \qquad \hat{R}_{\text{ref}} = \hat{c}_{p_{\text{ref}}}.$$

For compressible flows, the pressure reference is usually defined as

$$\hat{p}_{\text{ref}} = \hat{\rho}_{\text{ref}} \hat{v}_{\text{ref}}^2.$$

Typical reference values are $\hat{\rho}_{\text{ref}} = 1.2 \text{kg/m}^3$ which is approximately the density of air at room temperature or $\hat{v}_{\text{ref}}$ as the modulus of a reasonable reference velocity for the problem considered, for example the speed of an airplane. Regarding the reference length, there is no clear way of choosing this. Typically, this is the length of an object that crucially determines the flow, for example the diameter of a cylinder or the length of a plane.

Additionally, we need references for the physical parameters and constants $\hat{\mu}_{\text{ref}}$ and $\hat{\kappa}_{\text{ref}}$, as well as possibly reference values for the external forces. Reasonable values for air at room temperature and at sea level are $\hat{\mu}_{\text{ref}} = 18 \cdot 10^{-6} \frac{\text{kg}}{\text{m s}}$ and $\hat{\kappa}_{\text{ref}} = 25 \cdot 10^{-3} \frac{\text{kg m}}{\text{s}^3 \text{K}}$. For the nondimensional μ, the nondimensional Sutherland law is obtained from a nondimensionalization of (2.11) as

$$\mu = T^{\frac{3}{2}} \left(\frac{1 + Su}{T + Su} \right), \tag{2.13}$$

with Su the nondimensional Sutherland constant, which is $Su = \frac{110\text{K}}{\hat{T}_{\text{ref}}}$ for air. Again, a suitable value for $\hat{T}_{\text{ref}}$ is 273K.

The Reynolds and Prandtl number, as well as the parameter M are dimensionless quantities, given by:

$$Re = \frac{\hat{\rho}_{\text{ref}}\hat{v}_{\text{ref}}\hat{x}_{\text{ref}}}{\hat{\mu}_{\text{ref}}}, \qquad Pr = \frac{\hat{\mu}_{\text{ref}}\hat{c}_{p_{\text{ref}}}}{\hat{\kappa}_{\text{ref}}} \quad \text{and} \quad M = \frac{\hat{v}_{\text{ref}}}{\hat{c}_{\text{ref}}}. \qquad (2.14)$$

The Reynolds and the Prandtl numbers characterize important flow properties like the size of the boundary layer or if a flow is turbulent. Another important nondimensional quantity is the Mach number Ma, after the German engineer Ernst Mach, which is the quotient of the velocity and the speed of sound $\hat{c}$. The latter is given by

$$\hat{c} = \sqrt{\gamma \frac{\hat{p}}{\hat{\rho}}}. \qquad (2.15)$$

For Mach numbers near zero, compressible effects are typically absent and the flow can be viewed as incompressible. This is called the low Mach number regime. The nondimensional number M is related to the Mach number Ma via $M = \sqrt{\gamma}Ma$ and thus $M = \mathcal{O}_S(Ma)$.

All in all, in the three-dimensional case, we obtain the following set of equations for the nondimensional Navier-Stokes equations:

$$\partial_t \rho + \nabla \cdot \mathbf{m} = 0,$$

$$\partial_t m_i + \sum_{j=1}^{3} \partial_{x_j}(m_i v_j + p\delta_{ij}) = \frac{1}{Re}\sum_{j=1}^{3} \partial_{x_j} S_{ij} \qquad i = 1,2,3 \qquad (2.16)$$

$$\partial_t \rho E + \nabla \cdot (H\mathbf{m}) = \frac{1}{Re}\sum_{j=1}^{3} \partial_{x_j}\left(\sum_{i=1}^{3} S_{ij}v_i - \frac{W_j}{Pr}\right).$$

In short, using the vector of conservative variables $\mathbf{u} = (\rho, m_1, m_2, m_3, \rho E)^T$, the convective fluxes

$$\mathbf{f}_1^c(\mathbf{u}) = \begin{pmatrix} m_1 \\ m_1 v_1 + p \\ m_2 v_1 \\ m_3 v_1 \\ \rho H v_1 \end{pmatrix}, \quad \mathbf{f}_2^c(\mathbf{u}) = \begin{pmatrix} m_2 \\ m_1 v_2 \\ m_2 v_2 + p \\ m_3 v_2 \\ \rho H v_2 \end{pmatrix}, \quad \mathbf{f}_3^c(\mathbf{u}) = \begin{pmatrix} m_3 \\ m_1 v_3 \\ m_2 v_3 \\ m_3 v_3 + p \\ \rho H v_3 \end{pmatrix}, \quad (2.17)$$

and the viscous fluxes, where we emphasize the dependence on $\nabla \mathbf{u}$

$$\mathbf{f}_1^v(\mathbf{u}, \nabla \mathbf{u}) = \frac{1}{Re}\begin{pmatrix} 0 \\ S_{11} \\ S_{21} \\ S_{31} \\ \mathbf{S}_1 \cdot \mathbf{v} - \frac{W_1}{Pr} \end{pmatrix}, \quad \mathbf{f}_2^v(\mathbf{u}, \nabla \mathbf{u}) = \frac{1}{Re}\begin{pmatrix} 0 \\ S_{12} \\ S_{22} \\ S_{32} \\ \mathbf{S}_2 \cdot \mathbf{v} - \frac{W_2}{Pr} \end{pmatrix},$$

$$\mathbf{f}_3^v(\mathbf{u}, \nabla \mathbf{u}) = \frac{1}{Re}\begin{pmatrix} 0 \\ S_{13} \\ S_{23} \\ S_{33} \\ \mathbf{S}_3 \cdot \mathbf{v} - \frac{W_3}{Pr} \end{pmatrix} \tag{2.18}$$

these equations can be written as

$$\mathbf{u}_t + \partial_{x_1}\mathbf{f}_1^c(\mathbf{u}) + \partial_{x_2}\mathbf{f}_2^c(\mathbf{u}) + \partial_{x_3}\mathbf{f}_3^c(\mathbf{u}) = \partial_{x_1}\mathbf{f}_1^v(\mathbf{u}, \nabla \mathbf{u}) + \partial_{x_2}\mathbf{f}_2^v(\mathbf{u}, \nabla \mathbf{u})$$
$$+ \partial_{x_3}\mathbf{f}_3^v(\mathbf{u}, \nabla \mathbf{u}), \tag{2.19}$$

or in a more compact form:

$$\mathbf{u}_t + \nabla \cdot \mathbf{f}^c(\mathbf{u}) = \nabla \cdot \mathbf{f}^v(\mathbf{u}, \nabla \mathbf{u}). \tag{2.20}$$

These equations form a system of second-order partial differential equations of mixed hyperbolic-parabolic type. If only the steady state is considered, the equations are elliptic-parabolic.

Finally, we can write this in integral form as

$$\partial_{\hat{t}} \int_\Omega \mathbf{u}\,d\Omega + \int_{\partial\Omega} \mathbf{f}^c(\mathbf{u}) \cdot \mathbf{n}\,ds = \int_{\partial\Omega} \mathbf{f}^v(\mathbf{u}, \nabla \mathbf{u}) \cdot \mathbf{n}\,ds. \tag{2.21}$$

The conservative variables have the drawback that momentum density and energy density are hard to measure in reality. There, it is more common to have access to the so-called primitive variables (ρ, v_1, v_2, v_3, p), or sometimes with pressure replaced by temperature. Therefore, these are often used to visualize the solution or to characterize input data, instead of the conservative variables.

2.3   Source terms

If external forces or source terms are present, these will be modeled by additional terms on the appropriate right-hand sides:

$$\partial_{\hat{t}} \int_\Omega \mathbf{u}\,d\Omega + \int_{\partial\Omega} \mathbf{f}^c(\mathbf{u}) \cdot \mathbf{n}\,ds = \int_{\partial\Omega} \mathbf{f}^v(\mathbf{u}, \nabla \mathbf{u}) \cdot \mathbf{n}\,ds + \int_\Omega \mathbf{g}\,d\Omega.$$

An important example is gravity, which appears as a vector valued source in the momentum equation:

$$\mathbf{g} = (0,0,0,g,0)^T.$$

Another example would be a local heat source in the energy equation or the coriolis force. Additional nondimensional quantities appear in front of the source terms. For the gravitational source term, this is the Froude number

$$Fr = \frac{\hat{v}_{\text{ref}}}{\sqrt{\hat{x}_{\text{ref}}\hat{g}_{\text{ref}}}}. \tag{2.22}$$

2.4 Simplifications of the Navier-Stokes equations

The compressible Navier-Stokes equations are a very general model for compressible flow. Under certain assumptions, simpler models can be derived. The most important simplification of the Navier-Stokes equations are the incompressible Navier-Stokes equations modeling incompressible fluids. In fact, this model is used so widely that sometimes these equations are referred to as the Navier-Stokes equations, and the more general model of these equations is referred to as the compressible Navier-Stokes equations. They are obtained on assuming that density is constant along the trajectories of particles. In this way, the energy equation is no longer needed and the continuity equation simplifies to the so-called divergence constraint:

$$
\begin{aligned}
\nabla \cdot \mathbf{v} &= 0, \\
\mathbf{v}_t + \mathbf{v} \cdot \nabla \mathbf{v} + \frac{1}{\rho}\nabla p &= \frac{\mu}{\rho}\Delta \mathbf{v}.
\end{aligned}
\tag{2.23}
$$

Note that the density is not necessarily constant here. However, since this is a common modeling assumption, for example for water, the above equations are often called the incompressible Navier-Stokes equations with variable density; the above equations with the density hidden via a redefinition of the other terms are referred to as the incompressible Navier-Stokes equations.

There is a different way of deriving the incompressible Navier-Stokes equations with variable density, namely by looking at the limit $M \to 0$. Using formal asymptotic analysis, it can be shown that the compressible Navier-Stokes equations result in these in the limit.

2.5   The Euler equations

An important simplification of the Navier-Stokes equations is obtained, if the second-order terms (viscosity and heat conduction) are neglected or otherwise put, if the limit Reynolds number to infinity is considered. The resulting set of first-order partial differential equations are the so-called Euler equations:

$$\partial_t \hat{\rho} + \nabla \cdot \hat{\mathbf{m}} = 0,$$

$$\partial_t \hat{m}_i + \sum_{j=1}^{d} \partial_{x_j} (\hat{m}_i \hat{v}_j + \hat{p} \delta_{ij}) = 0, \qquad i = 1, ..., d \qquad (2.24)$$

$$\partial_t (\hat{\rho} \hat{E}) + \nabla \cdot (\hat{H} \hat{\mathbf{m}}) = 0.$$

Using the standard nondimensionalization, we obtain the following form:

$$\partial_t \rho + \nabla \cdot \mathbf{m} = 0,$$

$$\partial_t m_i + \sum_{j=1}^{d} \partial_{x_j} (m_i v_j + p \delta_{ij}) = 0, \qquad i = 1, ..., d \qquad (2.25)$$

$$\partial_t \rho E + \nabla \cdot (H \mathbf{m}) = 0.$$

As can be seen, no dimensionless reference numbers appear. With the vector of conservative variables $\mathbf{u} = (\rho, m_1, m_2, m_3, \rho E)^T$ and the convective fluxes $\mathbf{f}_1^c(\mathbf{u})$, $\mathbf{f}_2^c(\mathbf{u})$, and $\mathbf{f}_2^c(\mathbf{u})$ as before, these equations can be written as

$$\mathbf{u}_t + \partial_{x_1} \mathbf{f}_1^c(\mathbf{u}) + \partial_{x_2} \mathbf{f}_2^c(\mathbf{u}) + \partial_{x_3} \mathbf{f}_3^c(\mathbf{u}) = \mathbf{0}$$

or in more compact form as

$$\mathbf{u}_t + \nabla \cdot \mathbf{f}^c(\mathbf{u}) = \mathbf{0}. \qquad (2.26)$$

2.5.1   Properties of the Euler equations

First of all, the Euler equations are purely hyperbolic, which means that the eigenvalues of $\nabla \mathbf{f}^c \cdot \mathbf{n}$ are all real for any $\mathbf{n}$. In particular, they are given by

$$\lambda_{1/2} = v_n \pm c,$$

$$\lambda_{3,4,5} = v_n. \qquad (2.27)$$

Thus, the equations have all the properties of hyperbolic equations. In particular, when starting from nontrivial smooth data, the solution will be discontinuous after a finite time.

From (2.27), it can be seen that in multiple dimensions, one of the eigenvalues of the Euler equations is a multiple eigenvalue, which means that the Euler equations are not strictly hyperbolic. Furthermore, the number of positive and negative eigenvalues depends on the relation between normal velocity and the speed of sound. For $|v_n| > c$, all eigenvalues have the same sign (positive or negative), whereas for $|v_n| < c$, the eigenvalues have different signs. Furthermore, there will be no eigenvalues for $|v_n| = c$ and $v_n = 0$. Physically, this means that for $|v_n| < c$, information is transported in two directions, whereas for $|v_n| > c$, information is transported in one direction only. Alternatively, this can be formulated in terms of the Mach number. This is of particular interest for the reference Mach number M, since this tells us how the flow of information in most of the domain looks like. Thus, the regime $M < 1$ is called the *subsonic regime* and the regime $M > 1$ the *supersonic regime*. Finally, in the regime $M \in [0.8, 1]$, we typically have the situation in which there are locally subsonic and supersonic flows; this regime is called *transonic*.

2.6 Solution theory

Basic mathematical questions about an equation are: Is there a solution to the equation? Is the solution unique? Is the equation stable in some sense? For a number of special cases, exact solutions of the Navier-Stokes equations have been provided. Helmholtz managed to give results for the case of flow of zero vorticity, and Prandtl derived equations for the boundary layer that allow to derive exact solutions. Otherwise, there are no satisfactory answers to these questions. Roughly speaking, the existing results provide either long-time results for very strict conditions on the initial data, or short-time results for weak conditions on the initial data. For a review, we refer to Lions [177]. The analysis of the Euler equations is extremely difficult as well, and there is a lack of results available [57].

The problem starts with that the integral form still requires differentiability of the solutions in time, which conflicts with the possibility of moving shocks. Therefore, one typically works with the notion of a weak solution that does not require differentiability at all, but only integrability. A function $u \in L^\infty$ is called a weak solution of the conservation law $u_t + f(u)_x = 0$, if it satisfies

$$\int_0^\infty \int_{-\infty}^\infty [\phi_t u + \phi_x f(u)] dx dt + \int_{-\infty}^\infty \phi(x, 0) u(x, 0) dx = 0 \quad \forall \phi \in C_0^1. \quad (2.28)$$

Unfortunately, while solutions with shocks can be weak solutions, weak solutions are typically nonunique. At this point, the standard approach is to go back to the original modeling process and note that one fundamental first principle of thermodynamics has not been used in deriving the Navier-Stokes equations, namely the second law of thermodynamics. This states that entropy has to be nonincreasing (note that in physics, the convention is for entropy to be nondecreasing). If this is used as an additional constraint, entropy solutions can be defined that are unique in the case of one-dimensional hyperbolic conservation laws. This constraint can also be put into the definition of the solution. To this end, it is sufficient to assume the existence of a smooth and convex entropy function $\eta(u)$ and a smooth entropy flux $\psi(u)$. Furthermore, we assume that

$$\psi'(u) = \eta'(u)f'(u).$$

An example is the physical entropy of the Euler equations,

$$\hat{s} = \hat{c}_v \ln(\hat{p}/\hat{\rho}^{-\gamma}). \tag{2.29}$$

Then we define a *weak entropy solution* using this inequality

$$\int_0^\infty \int_{-\infty}^\infty [\phi_t \eta(u) + \phi_x \psi(u)] dx dt + \int_{-\infty}^\infty \phi(x,0)\eta(u(x,0)) dx \geq 0, \tag{2.30}$$

where, the test functions $\phi \in C_0^1$ additionally have to be positive.

Chiodaroli and colleagues proved that in multiple dimensions, there are cases with nonunique weak entropy solutions for the Euler equations [58]. This, together with strong numerical evidence, has given rise to the concept of measure valued solutions [92].

2.7   Boundary and initial conditions

Initially, at time $\hat{t}_0$, we have to prescribe values for all variables, where it does not matter whether we use the conservative variables or any other set, as long as there is a one to one mapping to the conservative variables. We typically use the primitive variables, as these can be measured quite easily, in contrast to the conservative variables. Further, if we restrict ourselves to a compact set $D \in \mathcal{U}$, we have to prescribe conditions for the solution on the boundary. This is necessary for numerical calculations and, therefore, D is also called the computational domain. The number of boundary conditions needed depends on the type of the boundary.

2.7.1   Initial conditions

At time $t = t_0$, we typically define the primitive values, such as a velocity v_0, a density ρ_0 and a pressure p_0. All other values like the energy and the momentum will be computed from these.

2.7.2 Fixed wall conditions

At the wall, no-slip conditions are the conditions to use for viscous flows; thus the velocity should vanish, that is, $\mathbf{v} = 0$. Regarding the temperature, we use either isothermal boundary conditions, where the temperature is prescribed or adiabatic boundary conditions, where the normal heat flux is set to zero.

Due to the hyperbolic nature of the Euler equations, there are fewer possible conditions at the wall. Therefore, the slip condition is employed, thus only the normal velocity should vanish, that is, $v_n = 0$.

2.7.3 Moving walls

At a moving wall, the condition is that the flow velocity has to be the same as the wall velocity $\dot{\mathbf{x}}$. This leads to the following equation:

$$\int_{\Omega} \mathbf{u}_t d\Omega + \int_{\partial\Omega} \mathbf{f}(\mathbf{u}) - \dot{\mathbf{x}} \cdot \mathbf{n} ds = \int_{\partial\Omega} \mathbf{f}^v(\mathbf{u}, \nabla\mathbf{u}) \cdot \mathbf{n} ds. \qquad (2.31)$$

2.7.4 Periodic boundaries

To test numerical methods, periodic boundary conditions can be useful. Given a set of points $\mathbf{x}_1$ on a boundary Γ_1, these are mapped to a different set of points $\mathbf{x}_2$ on a boundary Γ_2:

$$\mathbf{u}(\mathbf{x}_1) = \mathbf{u}(\mathbf{x}_2). \qquad (2.32)$$

2.7.5 Farfield boundaries

Finally, there are boundaries that are chosen purely for computational reasons, sometimes referred to as farfield boundaries. Here, the computational domain ends out of necessity for it to be finite, not because the physical domain would end. From a mathematical point of view, one question is which boundary conditions lead to a well-posed problem and how many boundary conditions can be posed in the first place [163]. For the Euler equations, the crucial property is the number of incoming waves, which can be determined using the theory of characteristics. This means that the sign of the eigenvalues (2.27) has to be determined and negative eigenvalues in normal direction correspond to incoming waves. As shown in [213] for the Navier-Stokes equations using the energy method, the number of boundary conditions there differs significantly. Note that the question is open, if these boundary conditions lead to the same solution on the small domain as for the Navier-Stokes equation on an unbounded domain. The number of boundary conditions that lead to a well-posed problem is shown in Table 2.1.

For the Navier-Stokes equations, a full set of boundary conditions has to be provided at both supersonic and subsonic inflows and one less at the

	Euler equations	Navier-Stokes equations
Supersonic inflow	5	5
Subsonic inflow	4	5
Supersonic outflow	0	4
Subsonic outflow	1	4

TABLE 2.1: Number of boundary conditions to be posed in three dimensions.

outflow. Intuitively, this can be explained through the continuity equation being a transport equation only, whereas the momentum and energy equations have a second-order term. For the Euler equations and subsonic flow at an inflow boundary, we have to specify four values, as we have one outgoing wave corresponding to the eigenvalue $v_n - c < 0$. At the outflow boundary, we have four outgoing waves and one incoming wave, which again corresponds to the eigenvalue $v_n - c$. For supersonic flow, we have only incoming waves at the inflow boundary, respectively no incoming waves, which means that we cannot prescribe anything there.

2.8   Boundary layers

Another important property of the Navier-Stokes equations is the presence of boundary layers [244]. Mathematically, this is due to the parabolic nature and therefore, boundary layers are not present in the Euler equations. We distinguish two types of boundary layers: the velocity boundary layer due to the slip condition, where the tangential velocity changes from zero to the free stream velocity and the thermal boundary layer in case of isothermal boundaries, where the temperature changes from the wall temperature to the free stream temperature. The thickness of the boundary layer is of the order $1/\text{Re}$, where the velocity boundary layer is a factor of Pr larger than the thermal one. This means that the higher the Reynolds number, the thinner the boundary layer and the steeper the gradients inside.

An important flow feature that boundary layers can develop is flow separation, where the boundary layer stops being attached to the body, typically by forming a separation bubble (see Figure 2.1).

2.9   Laminar and turbulent flows

When looking at low speed flow around an object, the observation is made that the flow is streamlined and mixing between neighboring flows is very

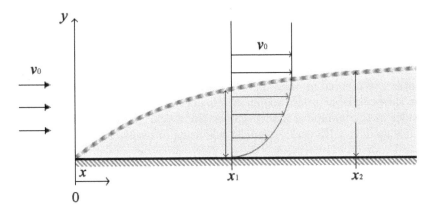

FIGURE 2.1: Boundary layer. (Credit: Tuso, CC-by-SA 3.0 (https://creativecommons.org/licenses/by-sa/3.0/). No changes were made.)

limited. However, when the speed is increased, then at some point the streamlines start to break, eddies appear, and neighboring flows mix significantly (see Figure 2.2). The first is called the *laminar* flow regime, whereas the other one is the *turbulent* flow regime. In fact, turbulent flows are chaotic in nature. As mentioned in the Introduction, we will consider almost only laminar flows.

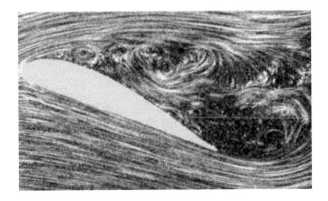

FIGURE 2.2: Turbulent flow. (Credit: Jaganath, CC-by-SA 3.0 (https://creativecommons.org/licenses/by-sa/3.0/). No changes were made.)

The same qualitative behavior can be observed dependend on the size of the object and the inverse of the viscosity. More precisely, for any object, there is a certain critical Reynolds number at which a laminar flow starts to become turbulent. The dependency on the inverse of the viscosity means air flows are typically turbulent; for example, the Reynolds number of a commercial airliner is between 10^6 and 10^8 for the A-380, whereas the critical Reynolds number is of the order 10^3. The exact determination of the critical Reynolds number is very difficult.

Simulating and understanding turbulent flows is still a very challenging problem. An important property of turbulent flows is that significant flow features are present at very different scales. The numerical method has to treat these different scales somehow. Essentially, there are two strategies to consider in the numerical model: Direct Numerical Simulation (DNS) or turbulence models. DNS uses extremely fine grids to resolve turbulent eddies directly. Unfortunately, resolving the smallest eddies, as shown by Kolmogorov, requires points on a scale of $Re^{9/4}$ and therefore, this approach is infeasible for practical applications even on modern supercomputers and more importantly, will remain to be so [204]. This leads to the alternative of turbulence models.

2.9.1   Turbulence models

A turbulence model—derived from the Navier-Stokes equations—tries to resolve only the larger eddies rather than smaller ones. To this end, the effect of small-scale eddies is incorporated using additional terms in the original equations with additional equations to describe the added terms. Examples for these approaches are the Reynolds Averaged Navier-Stokes equations (RANS) and the Large Eddy Simulation (LES).

2.9.1.1   RANS equations

The RANS equations represent the standard in industry for turbulence modeling. They are obtained by formally averaging the Navier-Stokes equations in a certain sense, an idea that goes back to Reynolds. Thus, every quantity is represented by a mean value plus a fluctuation:

$$\phi(x,t) = \overline{\phi}(x,t) + \phi^{'}(x,t).$$

For the definition of the mean value $\overline{\phi}(x,t)$, several possibilities exist, as long as the corresponding average of the fluctuation $\overline{\phi}^{'}(x,t)$ is zero. The Reynolds average is used for statistically steady turbulent flows and is given by averaging over a time interval that is significantly smaller than the time step, but large enough to integrate over small eddies

$$\overline{\phi}(x,t) = \frac{1}{T} \int_{-T/2}^{T/2} \phi(x, t + \tau)d\tau,$$

whereas the ensemble average is defined by

$$\overline{\phi}(x,t) = \lim_{N \to \infty} \frac{1}{N} \sum_{i=1}^{N} \phi_i,$$

where ϕ_i represents the result of an experiment to determine the quantity ϕ.

Furthermore, to avoid the computation of mean values of products, the Favre or density-weighted average is introduced:

$$\tilde{\phi} = \overline{\rho\phi}/\overline{\rho} \tag{2.33}$$

with a corresponding different fluctuation

$$\phi(x,t) = \tilde{\phi}(x,t) + \phi''(x,t).$$

When applying the averaging process to the Navier-Stokes equations, the continuity equation remains formally unchanged. In the momentum and energy equations, we obtain additional terms, the so-called Reynolds stresses $\mathbf{S}^R$ and the turbulent energy. We thus have the RANS equations (sometimes more correctly referred to as the Favre- and Reynolds-Averaged Navier-Stokes equations):

$$\partial_t \overline{\rho} + \nabla \cdot \overline{\rho}\tilde{\mathbf{v}} = 0,$$

$$\partial_t \overline{\rho}\tilde{\mathbf{v}} + \sum_{j=1}^{d} \partial_{x_j}(\overline{\rho}\tilde{v}_i\tilde{v}_j + \overline{p}\delta_{ij}) = \frac{1}{Re}\sum_{j=1}^{d}\partial_{x_j}\left(\tilde{S}_{ij} + S_{ij}^R\right), \quad i = 1, ..., d \tag{2.34}$$

$$\partial_t \overline{\rho}\tilde{E} + \nabla \cdot (\overline{\rho}\tilde{H}\tilde{v}_j) = \sum_{j=1}^{d}\partial_{x_j}\left(\left(\frac{1}{Re}S_{ij} - S_{ij}^R\right)v_i + \widetilde{S_{ij}v_i''}\right.$$
$$\left. - \overline{\rho v_j''} - \overline{\rho v_j''k} + \frac{\overline{W_j}}{RePr}\right).$$

The Reynolds stresses

$$S_{ij}^R = -\overline{\rho v_i'' v_j''}$$

and the turbulent energy

$$k = \frac{1}{2}\sum_{j=1}^{d}\overline{v_j' v_j'}$$

cannot be related to the unknowns of this equation. Therefore, they have to be described using experimentally validated turbulence models. These differ by the number of additional partial differential equations used to determine the Reynolds stresses and the turbulent energy. There are zero equation models, like the Baldwin-Lomax-model where just an algebraic equation is employed [7], one equation models like the Spallart-Allmaras model [257] or two equation models, as in the well-known k-ϵ-models [221].

2.9.1.2 Large Eddy Simulation

A more accurate approach is LES, which originally goes back to Smagorinsky, who developed it for meteorological computations [250]. This time, a filtering process with a filter G is used:

$$\overline{\phi(\mathbf{x}_0, t)} = \int_\Omega \phi(\mathbf{x}, t)G(\mathbf{x}_0, \mathbf{x}, \Delta)d\Omega.$$

This gives rise to a decomposition $\phi = \bar{\phi} + \phi'$, again with the property $\overline{\phi'} = 0$. However, as opposed to Reynolds averaging, this time no mixed terms disappear. The benefit arises in the modeling of these terms: If the filter is chosen fine enough, the filtered quantity ϕ' represents small eddies, which have a more homogenous structure than larger turbulent eddies. This homogeneity makes the modeling of the influence of the mixed terms easier than for the RANS equations. If the boundary layer is to be resolved, LES requires a resolution on the order of $Re^{9/5}$, compared to $Re^{9/4}$ for DNS. If a lower resolution is chosen, then the modeling error is bound to be larger than for the RANS equations. This is a big if. Essentially, LES computations are only feasible using massively parallel high-order methods.

With regards to filters, this can be done explicitly, for example using a box filter or a Gaussian filter. Alternatively, the numerical method on a specific grid implicitly defines a filter. The first approach has the advantage that the modeling error is smaller, however coming at the price of a more costly method and a larger implementation effort required.

For the compressible Navier-Stokes equations, the filtering is combined with Favre averaging (2.33). This is to avoid the need to model mixed terms in the continuity equation. We thus arrive at the Favre-filtered Navier-Stokes equations:

$$\partial_t \bar{\rho} + \nabla \cdot \bar{\rho} \tilde{\mathbf{v}} = 0,$$

$$\partial_t \bar{\rho} \tilde{\mathbf{v}} + \sum_{j=1}^{d} \partial_{x_j} \left(\bar{\rho} \tilde{v}_i \tilde{v}_j + \bar{p} \delta_{ij} \right) = \frac{1}{Re} \sum_{j=1}^{d} \partial_{x_j} \tilde{S}_{ij} + \sum_{j=1}^{d} \partial_{x_j} S_{ij}^{\text{SGS}}, \quad i = 1, ..., d,$$

$$(2.35)$$

$$\partial_t \bar{\rho} \tilde{E} + \nabla \cdot \left((\bar{\rho} \tilde{E} + \bar{p} \delta_{ij}) \tilde{v}_j \right) = \frac{1}{Re} \sum_{j=1}^{d} \partial_{x_j} \left(\tilde{S}_{ij} \tilde{v}_i + \frac{\overline{W}_j}{Pr} \right)$$

$$+ \sum_{j=1}^{d} \partial_{x_j} S_{ij}^{\text{SGS}} \tilde{v}_i + \sum_{j=1}^{d} \partial_{x_j} W_j^{\text{SGS}}.$$

The quantities S_{ij}^{SGS} and W_j^{SGS} are called subgrid scale stresses and subgrid scale heat flux, respectively, and need to be modeled. They arise from interaction between filtered and unfiltered quantities. The filtered stresses are given by

$$\tilde{S}_{ij} = \mu(\tilde{T}) \left((\partial_{x_j} \tilde{v}_i + \partial_{x_i} \tilde{v}_j) - \frac{2}{3} \delta_{ij} \nabla \cdot \tilde{\mathbf{v}} \right).$$

Compared to the turbulence models in the RANS equations, subgrid scale models have the advantage of having significantly fewer parameters to tune, and the choice of these parameters has a stronger physical backing. Possibilities are mainly the Smagorinsky model, which features a single parameter and dynamic SGS models, which replace the Smagorinsky constant with a time- and space-dependent function.

An important idea is to not use a formal subgrid scale model at all, which corresponds to using the numerical method as is for the compressible Navier-Stokes equations without any additional modeling steps. In this case, the subgrid scale terms are given implicitly by the numerical dissipation, which is why this approach is called implicit LES or ILES. If the discretization is fine enough, this approach can work very well [86].

2.9.1.3   Detached Eddy Simulation

Finally, there is the detached eddy simulation (DES) of Spallart [258]. This is a mixture between RANS and LES, where a RANS model is used in the boundary layer, whereas an LES model is used for regions with flow separation. Since then, several improved variants have been suggested, for example delayed detached eddy simulation (DDES).

Chapter 3

The space discretization

As discussed in the Introduction, we now seek approximate solutions to the continuous equations, where we will employ the methods of lines approach, which means that we first discretize in space to transform the equations into ordinary differential equations and then discretize in time. Regarding space discretizations, there is a plethora of methods available. The oldest methods are finite difference schemes that approximate spatial derivatives by finite differences, but these become very complicated for complex geometries or high orders. Furthermore, the straightforward methods have problems to mimic core properties of the exact solution like conservation or nonoscillatory behavior. While in recent years, a number of interesting new methods have been suggested, it remains to be seen whether these methods are competitive, respectively where their niches lie. In the world of elliptic and parabolic problems, finite element discretizations are standard. These use a set of basis functions to represent the approximate solution and then seek the best approximation defined by a Galerkin condition. For elliptic and parabolic problems, these methods are backed by extensive results from functional analysis, which make them very powerful. However, they have problems with convection-dominated problems, where additional stabilization terms are needed. This is an active field of research in particular for the incompressible Navier-Stokes equations, but the use of finite element methods for compressible flows is currently very limited.

The methods that are standard in industry and academia are finite-volume schemes. These use the integral form of a conservation law and consider the change of the mean of a conservative quantity in a cell via fluxes over the boundaries of the cell. Thus, the methods inherently conserve these quantities and furthermore can be made to satisfy additional properties of the exact solutions. Finite volume methods will be discussed in section 3.2. A problem of finite volume schemes is their extension to orders above two. A way of achieving this are discontinuous Galerkin methods that use ideas originally developed in the finite element world. These will be considered in section 3.7.

3.1   Structured and unstructured grids

Before we describe the space discretization, we will discuss different types of grids, namely structured and unstructured grids (see Fig. 3.1). The former grids have a certain regular structure, whereas unstructured grids do not. In particular, cartesian grids are structured, and also O- and C-type grids, which are obtained by mapping a cartesian grid using a Möbius transformation. The main advantage of structured grids is that the data structure is simpler; for example, the number of neighbors of a grid cell is a priori known, and thus the algorithm is easier to implement, as well as requiring less memory. Furthermore, the simpler geometric structure also leads to easier analysis of numerical methods, which often translates in more robustness and speed.

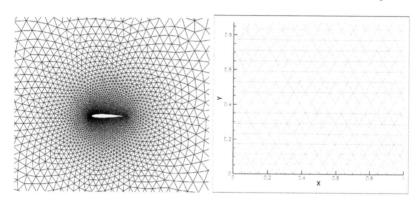

FIGURE 3.1: Example of unstructured (left) and structured (right) triangular grids

On the other hand, the main advantage of unstructured grids is that they are geometrically much more flexible, which allows for a better resolution of arbitrary geometries.

At a fixed wall, the representation of the boundary will typically be inexact, which will introduce additional errors. For a second-order method, a piecewise linear approximation of the boundary is sufficient. This implies a polygonal mesh. However, this will reduce the order for methods of higher order. There, a curved representation of the boundary is desired, which significantly increases the complexity of the grid representation.

When generating a grid for the solution of the Navier-Stokes equations, an important feature to consider is the boundary layer. Since there are huge gradients in normal direction, but not in tangential direction, it is useful to use cells in the boundary layer that have a high aspect ratio; the higher the Reynolds number, the higher the aspect ratio. Furthermore, to avoid cells with extreme angles, the boundary layer is often discretized using a structured grid.

Grid generation continues to be a bottle-neck in CFD calculations, as automatic procedures for doing so are missing. Possible codes are for example the commercial software CENTAUR [56] or the open source Gmsh [105]. Tools that help with curved boundaries are the HOPR preprocessor [132] or Seeder [159].

3.2   Finite volume methods

The equations we are trying to solve are so-called conservation, respectively balance laws. For these, finite volume methods are the most natural to use. Basis for those is the integral form of the equation. An obvious advantage of this formulation is that discontinuous solutions of some regularity are admissible. This is favorable for nonlinear hyperbolic equations, because shocks are a common feature of their solutions. We keep the presentation brief and refer the interested reader for more information to the excellent textbooks [107, 135, 136, 174, 175] and the more concise treatises [208, 11]. We start the derivation with the integral form (2.21):

$$\frac{d}{dt} \int_\Omega \mathbf{u}(\mathbf{x}, t) \, d\Omega + \int_{\partial\Omega} \mathbf{f}(\mathbf{u}(\mathbf{x}, t)) \cdot \mathbf{n} \, ds = \int_\Omega \mathbf{g}(\mathbf{x}, t, \mathbf{u}(\mathbf{x}, t)) \, d\Omega. \qquad (3.1)$$

Here, Ω is a so-called control volume or cell with outer normal unit vector $\mathbf{n}$. The only condition we need to put on Ω for the derivation to work is that it satisfies the assumption of the divergence theorem, meaning that is has a Lipschitz continuous boundary. However, we now assume that all control volumes have polygonal boundaries. This is not a severe restriction and allows for a huge amount of flexibility in grid generation, which is another advantage of finite volume schemes. Thus we decompose the computational domain D into a finite number of polygons in two dimensions, respectively polygonally bounded sets in three dimensions.

We denote the ith cell by Ω_i and its volume by $|\Omega_i|$. Edges will be called e with the edge between Ω_i and Ω_j being e_{ij} with length $|e_{ij}|$, whereby we use the same notation for surfaces in 3D. Furthermore, we denote the set of indices of the cells neighboring cell i with $N(i)$. We can therefore rewrite (3.1) for each cell as

$$\frac{d}{dt} \int_{\Omega_i} \mathbf{u}(\mathbf{x}, t) \, d\Omega + \sum_{j \in N(i)} \int_{e_{ij}} \mathbf{f}(\mathbf{u}(\mathbf{x}, t)) \cdot \mathbf{n} \, ds = \int_{\Omega_i} \mathbf{g}(\mathbf{x}, t, \mathbf{u}(\mathbf{x}, t)) \, d\Omega. \quad (3.2)$$

The key step toward a numerical method is now to consider the mean value

$$u_i(t) := \frac{1}{|\Omega_i|} \int_{\Omega_i} u(x, t) d\Omega$$

of $u(x, t)$ in every cell Ω_i and to use this to approximate the solution in the cell. Under the condition that Ω_i does not change with time, we obtain an evolution equation for the mean value in a cell:

$$\frac{d}{dt} u_i(t) = -\frac{1}{|\Omega_i|} \sum_{j \in N(i)} \int_{e_{ij}} f(u(x, t)) \cdot n \, ds + \frac{1}{|\Omega_i|} \int_{\Omega_i} g(x, t, u(x, t)) \, d\Omega.$$

(3.3)

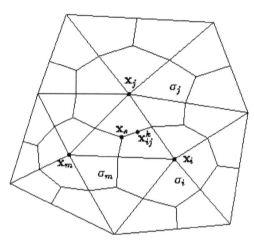

FIGURE 3.2: Primary triangular grid and corresponding dual grid for a cell-vertex scheme. (Credit: A. Meister.)

We will now distinguish two types of schemes, namely *cell-centered* and *cell-vertex* schemes. For cell-centered schemes, the grid is used as generated, meaning that the unknowns do not correspond to vertices, but to the cells of the grid. This is a very common type of scheme, in particular for the Euler equations. In the case of cell-vertex schemes, the unknowns are located in the vertices of the original (primary) grid. This is possible after creating a dual grid with dual control volumes, by computing the barycenter of the original grid and connecting this to the midpoints of the edges of the containing primary cell (see Fig. 3.2). The finite volume method then acts on the dual cells.

For cartesian grids, there is no big difference between the two approaches, since the dual cells are quadrilaterals as well. However, when the primary cells are triangles or tetrahedrons, then the dual cells are complicated polygons and the difference is significant. For example, the angles between two edges in the dual cells are larger, which later implies a smaller discretization error and thus the possibility of using larger cells. However, the main purpose of these schemes is to make the calculation of velocity gradients on unstructured grids easier, which is important in the context of Navier-Stokes equations. This will be discussed in section 3.3.3.

3.3 The line integrals and numerical flux functions

In equation (3.2), line integrals of the flux along the edges appear. A numerical method thus needs a mean to compute them. On the edge though, the numerical solution is usually discontinuous, because it consists of the mean values in the cells. Therefore, a numerical flux function is required. This takes states from the left and the right side of the edge and approximates the exact flux based on these states. For now, we assume that the states are the mean values in the respective cells, but later when considering higher-order methods in section 3.6, other values are possible. The straightforward way to define a numerical flux function would be to simply use the average of the physical fluxes from the left and the right. However, this leads to an unconditionally unstable scheme and therefore, additional stabilizing terms are needed.

A numerical flux function $\mathbf{f}^N(\mathbf{u}_L, \mathbf{u}_R; \mathbf{n})$ is called consistent, if it is Lipschitz continuous in the first two arguments and if $\mathbf{f}^N(\mathbf{u}, \mathbf{u}; \mathbf{n}) = \mathbf{f}(\mathbf{u}) \cdot \mathbf{n}$. This essentially means that if we refine the discretization, the numerical flux function will better approximate the physical flux.

We furthermore call a numerical flux function $\mathbf{f}^N$ *rotationally invariant* if a rotation of the coordinates does not change the flux. The physical flux has this property; therefore, it is reasonable to require this also of the numerical flux. More precisely, this property is given in the two-dimensional case by

$$\mathbf{T}^{-1}(\mathbf{n})\mathbf{f}^N(\mathbf{T}(\mathbf{n})\mathbf{u}_L, \mathbf{T}(\mathbf{n})\mathbf{u}_R; (1,0)^T) = \mathbf{f}^N(\mathbf{u}_L, \mathbf{u}_R; \mathbf{n}).$$

In two dimensions, the matrix $\mathbf{T}$ is given by

$$\mathbf{T}(\mathbf{n}) = \begin{pmatrix} 1 & & \\ & n_1 & n_2 \\ & -n_2 & n_1 \\ & & & 1 \end{pmatrix}.$$

The rotation matrix in three dimensions is significantly more complicated. This property can be made use of in the code, as it allows to assume that the input of the numerical flux function is aligned in normal direction and therefore, it is sufficient to define the numerical flux functions only for $\mathbf{n} = (1, 0, 0)^T$.

In the derivation so far, we have not used any properties of the specific equation, except for the rotational invariance, which is a property that most

physical systems have. This means that the flux functions contain the important information about the physics of the equations considered and that for
finite volume methods, special care should be taken in designing them. There
is a significant advantage to this, namely that given a finite volume code to
solve a certain equation, it is rather straightforward to adjust the code to solve
a different conservation law.

Employing any flux function, we can now approximate the line integrals
by a quadrature formula. A Gaussian quadrature rule with one Gauss point
in the middle of the edge already achieves second-order accuracy, which is
sufficient for the finite volume scheme used here. The approximation obtained
is therefore

$$\int_{e_{ij}} \mathbf{f}(\mathbf{u}(\mathbf{x},t)) \cdot \mathbf{n} \, ds \approx |e_{ij}| \mathbf{T}^{-1}(\mathbf{n}_{ij}) \mathbf{f}^N(\mathbf{T}(\mathbf{n}_{ij})\mathbf{u}_i, \mathbf{T}(\mathbf{n}_{ij})\mathbf{u}_j; (1,0,0)^T).$$

Thus, we can write down a semidiscrete form of the conservation law without source terms, namely a finite dimensional nonlinear system of ordinary
differential equations. This way of treating a partial differential equation is
also called the method of lines approach. For each single cell, this differential
equation can be written as:

$$\frac{d}{dt}\mathbf{u}_i(t) + \frac{1}{|\Omega_i|} \sum_{j \in N(i)} |e_{ij}| \mathbf{T}^{-1}(\mathbf{n}_{ij}) \mathbf{f}^N(\mathbf{T}(\mathbf{n}_{ij})\mathbf{u}_i, \mathbf{T}(\mathbf{n}_{ij})\mathbf{u}_j; (1,0,0)^T) = \mathbf{0},$$

(3.4)

where the input of the flux function are the states on the left-hand side and
right-hand side of the edge, respectively.

To implement a finite volume method, the above equation suggests to do
a loop over the cells, which represent the unknowns. However, then all fluxes
would be computed twice and conservation is not explicitly respected in the
code. Therefore, one does instead a loop over edges and computes the flux
over that edge. This is then added and subtracted, respectively, depending on
the normal vector, from the solution in the two neighboring cells. The data
structure that represents the mesh thus stores cells and their volume, edges
and their length, and normal vectors and for each edge what the neighboring
volumes are.

3.3.1   Discretization of the inviscid fluxes

Numerical flux functions for the Euler equations are numerous and will be
presented here only in brief. For more detailed information, consult [136, 277]
and the references therein. There is a long history, starting in a way with
Godunov's method in the 1950s. He had the idea of interpreting the discontinuous solution at the edge as a Riemann problem. There, we have an infinite
domain with constant data, except for one discontinuity. For many equations,
as well as the Euler equations, this problem admits a unique solution, which

can be constructed, giving rise to a numerical method. As it turns out, the construction of the exact solution is both extremely expensive and total overkill, which gave rise to approximate Riemann solvers, where only an approximate solution is the constructed. Examples for this are the Roe scheme, as well as the HLL family of schemes, of which HLLC is presented in detail in the next section.

Riemann problems are not the only idea used to define numerical flux functions. The AUSM family, of which AUSMDV is explained later, uses a splitting in a pressure and a convective part. Another possibility is to use an averaging of the physical fluxes on the left and the right, and a stabilizing diffusion term. Along this line, Jameson developed the JST scheme, as well as the CUSP and SLIP methods [143]. In the context of the later explained DG methods, the extremely simple Rusanov flux is also employed, which is of the same type. Van Leers flux vector splitting [286] is not a good method to compute flows with strong boundary layers.

In the following descriptions of numerical flux functions, we will assume that they have the vectors $\mathbf{u}_L$ and $\mathbf{u}_R$ as input, with their components indexed by L and R, respectively.

3.3.1.1   HLLC

The HLLC flux of Toro is from the class of approximate Riemann solvers that consider a Riemann problem and solve that approximately. HLLC tries to capture all waves in the problem to then integrate an approximated linearized problem (see Fig. 3.3). It is an extension of the HLL Riemann solver of Harten,

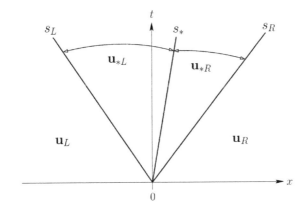

FIGURE 3.3: The solution of the linearized Riemann problem used in HLLC: Four different states $\mathbf{u}_L$, $\mathbf{u}_R$, $\mathbf{u}_{*L}$, and $\mathbf{u}_{*R}$ are separated by two shock waves with speeds s_L and s_R and a contact discontinuity of speed s_*. (Credit: V. Straub.)

Hyman, and van Leer, but additionally includes a contact discontinuity, thus the C. To this end, the speeds s_L and s_R of the left and right going shock are determined and the speed of the contact discontinuity is approximated as s^*. With these, the flux is determined as

$$\mathbf{f}(\mathbf{u}_L, \mathbf{u}_R; \mathbf{n}) = \begin{cases} \mathbf{f}(\mathbf{u}_L), & 0 \leq s_L, \\ \mathbf{f}_L^*, & s_L \leq 0 \leq s^*, \\ \mathbf{f}_R^*, & s^* \leq 0 \leq s_R, \\ \mathbf{f}(\mathbf{u}_R), & s_R \geq 0, \end{cases} \tag{3.5}$$

where

$$\mathbf{f}_L^* = \mathbf{f}(\mathbf{u}_L) + s_L(\mathbf{u}_{*L} - \mathbf{u}_L)$$

and analogously,

$$\mathbf{f}_R^* = \mathbf{f}(\mathbf{u}_R) + s_R(\mathbf{u}_{*R} - \mathbf{u}_R).$$

Here, the intermediate states $\mathbf{u}_{*K}$ are given by

$$\mathbf{u}_{*K} = \rho_K \left(\frac{s_K - v_{1_K}}{s_K - s^*} \right) \begin{pmatrix} 1 \\ s^* \\ v_{2_K} \\ v_{3_K} \\ \frac{E_K}{\rho_K} + (s^* - v_{1_K})\left(s^* - \frac{p_K}{\rho_K(s_K - v_{1_K})}\right) \end{pmatrix}.$$

The intermediate speed is given by

$$s^* = \frac{1}{2}\{v_1\} + \frac{1}{2}(p_L - p_R)/(\hat{\rho}\hat{c}),$$

where $\{v_1\}$ is the left-right averaged quantity and

$$\hat{\rho} = \frac{1}{2}\{\rho\}, \quad \hat{c} = \frac{1}{2}\{c\}.$$

Furthermore, the speeds of the left and right going shock are found as

$$s_L = v_{1_L} - c_L q_L, \quad s_R = v_{1_R} + c_R q_R.$$

Finally, the wave speeds q_K for $K = L, R$ are obtained from

$$q_k = \begin{cases} 1, & p^* \leq p_K \\ \sqrt{1 + \frac{\gamma+1}{\gamma}\left(\frac{p^*}{p_K} - 1\right)}, & p^* > p_K \end{cases}$$

with

$$p^* = \frac{1}{2}\{p\} + \frac{1}{2}(v_{1_L} - v_{1_R})/(\hat{\rho}\hat{c}).$$

3.3.1.2 AUSMDV

By contrast to many other numerical flux functions like HLLC, the idea behind the fluxes of the AUSM family is not to approximate the Riemann problem and integrate that, but to approximate the flux directly. To this end, the flux is written as the sum of a pressure term and a flow term. The precise definition of these then determines the exact AUSM-type scheme. The AUS-MDV flux [296] is actually a combination of several fluxes, namely AUSMD, AUSMV, the flux function of Hänel and a shock fix. For AUSMD, we have

$$
\mathbf{f}^{AUSMD}(\mathbf{u}_L, \mathbf{u}_R; \mathbf{n}) = \frac{1}{2}[(\rho v_n)_{1/2}(\Psi_L + \Psi_R) + |(\rho v_n)_{1/2}|(\Psi_L - \Psi_R)] + \mathbf{p}_{1/2},
$$
(3.6)

with

$$
\Psi = (1, v_1, v_2, v_3, H)^T \text{ and } \mathbf{p}_{1/2} = (0, p_{1/2}, 0, 0,, 0)^T.
$$

Here,

$$
p_{1/2} = p_L^+ + p_R^-,
$$

with

$$
p_{L/R}^{\pm} = \begin{cases} \frac{1}{4}p_{L/R}(M_{L/R} \pm 1)^2(2 \mp M_{L/R}) & \text{if } |M_{L/R}| \leq 1, \\ \frac{1}{2}p_{L/R}\frac{v_{n,L/R} \pm v_{n,L/R}}{v_{n,L/R}} & \text{else.} \end{cases}
$$

The local Mach numbers are defined as

$$
M_{L,R} = \frac{v_{n,L/R}}{c_m},
$$

with $c_m = \max\{c_L, c_R\}$. Furthermore, the mass flux is given by

$$
(\rho v_n)_{1/2} = v_{n,L}^+ \rho_L + v_{n,R}^- \rho_R,
$$

with

$$
v_{n,L/R}^{\pm} = \begin{cases} \alpha_{L,R}\left(\pm\frac{(v_{n,L/R} \pm c_m)^2}{4c_m} - \frac{v_{n,L/R} \pm |v_{n,L/R}|}{2}\right) + \frac{v_{n,L/R} \pm |v_{n,L/R}|}{2} & \text{if } |M_{L/R}| \leq 1, \\ \frac{v_{n,L/R} \pm v_{n,L/R}}{2} & \text{else.} \end{cases}
$$

The two factors α_L and α_R are defined as

$$
\alpha_L = \frac{2(p/\rho)_L}{(p/\rho)_L + (p/\rho)_R} \text{ and } \alpha_R = \frac{2(p/\rho)_R}{(p/\rho)_L + (p/\rho)_R}.
$$

As the second ingredient of AUSMDV, the AUSMV flux is identical to the AUSMD flux, except that the first momentum component is changed to

$$
f_{m_1}^{AUSMV}(\mathbf{u}_L, \mathbf{u}_R; \mathbf{n}) = v_{n,L}^+(\rho v_n)_L + v_{n,R}^-(\rho v_n)_R + p_{1/2}
$$
(3.7)

with $v_{n,L}^+$ and $v_{n,R}^-$ defined as above.

The point is that AUSMD and AUSMV have different behavior at shocks;

AUSMD causes oscillations at these. On the other hand, AUSMV produces wiggles in the velocity field at contact discontinuities. Therefore, the two are combined using a gradient-based sensor s:

$$\mathbf{f}^{AUSMD+V} = \left(\frac{1}{2}+s\right)\mathbf{f}^{AUSMD}(\mathbf{u}_L,\mathbf{u}_R;\mathbf{n}) + \left(\frac{1}{2}-s\right)\mathbf{f}^{AUSMV}(\mathbf{u}_L,\mathbf{u}_R;\mathbf{n}).$$
(3.8)

The sensor s depends on the pressure gradient in that

$$s = \frac{1}{2}\min\left\{1, K\frac{|p_R - p_L|}{\min\{p_R, p_R\}}\right\},$$

where we chose $K = 10$.

Now, the AUSMD+V still has problems at shocks, which is why the Hänel flux is used to increase damping. This is given by

$$\mathbf{f}^{Haenel}(\mathbf{u}_L,\mathbf{u}_R;\mathbf{n}) = v_{n,L}^{+}\rho_L\boldsymbol{\Psi}_L + v_{n,R}^{-}\rho_R\boldsymbol{\Psi}_R + \mathbf{p}_{1/2}, \qquad (3.9)$$

where everything is defined as above, except that $\alpha_L = \alpha_R = 1$ and c_m is replaced by c_L and c_R, respectively.

Furthermore, a damping function $\mathbf{f}^D$ is used at the sonic point to make the flux function continuous there. Using the detectors

$$A = ((v_{n,L} - c_L) < 0)\,\&\,(v_{n,R} - c_R) > 0)$$

and

$$B = ((v_{n,L} + c_L) < 0)\,\&\,(v_{n,R} + c_R) > 0),$$

as well as the constant $C = 0.125$, we define the damping function as

$$\mathbf{f}^D = \begin{cases} C((v_{n,L} - c_L) - (v_{n,R} - c_R))((\rho\Phi)_L) - (\rho\Phi)_R), & \text{if } A\,\&\,\neg B, \\ C((v_{n,L} + c_L) - (v_{n,R} + c_R))((\rho\Phi)_L) - (\rho\Phi)_R), & \text{if } \neg A\,\&\,B, \\ \mathbf{0}, & \text{else.} \end{cases}$$
(3.10)

Finally, we obtain

$$\mathbf{f}^{AUSMDV} = (1 - \delta_{2,S_L+S_R})(\mathbf{f}^D + \mathbf{f}^{AUSMD+V}) + (\delta_{2,S_L+S_R})\mathbf{f}^{Haenel}, \quad (3.11)$$

where

$$S_K = \begin{cases} 1, & \text{if } \exists j \in N(i) \quad \text{with } (v_{n_i} - c_i) > 0\,\&\,(v_{n_j} - c_j) < 0 \\ & \qquad\qquad\text{or } (v_{n_i} + c_i) > 0\,\&\,(v_{n_j} + c_j) < 0, \\ 0, & \text{else,} \end{cases}$$

is again a detector for the sonic point near shocks.

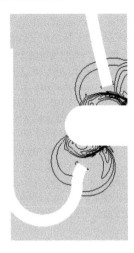

FIGURE 3.4: A flow is blown from the nozzles at Mach 0.01. Shown are isolines of pressure when using AUSMDV (left) and L^2Roe (right). While both solutions look reasonable, the pressure fluctuations in the AUSM solution are too high by a factor of 4.

3.3.2 Low Mach numbers

In the case of low Mach numbers, which means $M < 0.1$, care has to be taken, since all standard flux functions produce wrong solutions on reasonable grids, in particular in the pressure field. Spatial fluctuations in the pressure are of the order $\mathcal{O}(M^2)$, as M goes to zero, as an asymptotic analysis of the continuous equations [118] shows. In Figure 3.4, we show pressure isolines for a flow exiting two nozzles at Mach 0.01 and getting reflected at a flanged shaft (see section 8.7.1 about the underlying problem and 7.4 for the numerical experiment). On the left, AUSMDV is used as flux function, whereas on the right it is the method L^2Roe discussed below. As can be proven by discrete asymptotic analysis, AUSMDV produces pressure fluctuations on the order of $\mathcal{O}(M)$. In this case, they are four times higher than the other ones. Thus, the solution is just wrong.

From a historical point of view, the reason for the failure of standard methods is that for decades, the most important applications in CFD were hypersonic and supersonic flows and thus, the methods were designed for the resolution of strong shocks. The mathematical reason are jumps in the discrete normal velocity, as was shown by analyzing Riemann problems in [117]. Furthermore, as shown in [236], this problem does not exist on triangular grids. Since these jumps are a result of the discretization, they can in theory be mitigated by refining the grid or increasing the order. However, this is impractical, as then a constant discretization error requires increasingly finer meshes with decreasing Mach number.

One solution is the preconditioning technique of Guillard and Viozat [118]. However, this turns out to be unacceptably inefficient if combined with explicit time integration, as was shown by von Neumann stability analysis in [39] and later for a more general case in [73]. There, it is shown that the time step size has to decrease asymptotically with the Mach number squared. Another suggestion is the AUSM+up flux of Liou [179], which also solves the accuracy problem. However, this method has a number of parameters which are difficult to tune; though stable, this method shows a similar dependency of the solution on the time step as the preconditioning technique of Guillard and Viozat.

Therefore, more recent fixes focus on the decrease of the jumps in the normal component of the velocity at cell interfaces. For example, Rieper suggests the method LMRoe [235], a modification of Roe's flux function, where the jump Δv_n in the normal component across the interface is replaced with

$$\Delta v_n = \min(1, \tilde{M})\Delta v_n \qquad (3.12)$$

and the local Mach number $\tilde{M}$ is given by

$$\tilde{M} = \frac{|v_{n_{ij}}| + |v_{t_{ij}}|}{c_{ij}}$$

and all quantities are supposed to be Roe-averaged velocities, which means that we consider

$$\phi_{ij} = \frac{\sqrt{\rho_i}\phi_i + \sqrt{\rho_j}\phi_j}{\sqrt{\rho_i} + \sqrt{\rho_j}}.$$

For this fix, Rieper proves by asymptotic analysis that the resulting method leads to correctly scaled pressure and by von Neumann stability analysis that the time step scales with $\mathcal{O}(M)$.

While decreasing the jump on normal velocity is crucial, there is no principal reason not to reduce the jumps in the tangential velocity components. This was suggested first by Thornber et al. However, they modify the velocities directly, thereby changing the evaluation of the rest of the numerical flux function as well [274]. Therefore, it is proposed in [217] to apply the scaling (3.12) to all velocity jumps. There, it is also shown using asymptotic analysis that this reduces the numerical dissipation. The resulting scheme is called L^2Roe and shown to be particularly useful for low Mach turbulent flows, where a low numerical dissipation is crucial. This method is the one used in Figure 3.4 (right).

Even this scheme can fail for very low Mach numbers, where schemes as discussed in [202, 10, 49] are to be preferred.

3.3.3   Discretization of the viscous fluxes

The viscous fluxes (2.18) are easier to discretize, because, due to the parabolic nature of the second-order terms, central differences do not lead to an unconditionally unstable scheme. It is therefore possible to base the evaluation of the viscous fluxes at an edge on simple averages over the neighboring cells:

$$\phi_{ij} = \frac{\phi_i + \phi_j}{2}.$$

Here, ϕ corresponds to one of the needed quantities in the viscous fluxes, namely the viscosity ν, a velocity component v_i, or the thermal conductivity κ.

Furthermore, gradients of velocity and temperature need to be computed on the edge. Here, the naive implementation using arithmetic means unfortunately leads to a decoupling of neighboring cells, leading to the so-called checker board effect. Therefore, the coupling between cells needs to be recovered in some way.

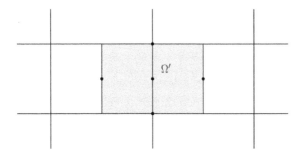

FIGURE 3.5: The dual cell Ω' used to compute the gradients for a structured cell-centered method. (Credit: V. Straub.)

In the context of a structured cell-centered finite-volume method, this is done by introducing a dual cell Ω' around the edge point where we want to compute the derivative (see Figure 3.5). Now we assume that the values of the derivatives at the cell centers are given. This is for example the case when a higher-order finite volume method is used, as explained later in section 3.6.2. The values in the other points are obtained using arithmetic means over all neighboring cells. To obtain the partial derivatives at the edge point, Green's theorem can be employed. We have:

$$\int_\Omega \frac{\partial \phi}{\partial x} d\Omega = \int_{\partial\Omega} \phi dy \quad \text{and} \quad \int_\Omega \frac{\partial \phi}{\partial y} d\Omega = \int_{\partial\Omega} \phi dx.$$

Computing the left-hand side and approximating the surface integrals using second-order Gaussian quadrature, we obtain for the domain Ω' in Figure 3.5:

$$\frac{\partial \phi}{\partial x} = \frac{1}{|\Omega'|} \int_{\partial\Omega'} \phi dy \approx \frac{1}{|\Omega'|} \sum_{e_m \in \partial\Omega'} \phi_m |e_m| n_{m_x}. \tag{3.13}$$

As an example in two dimensions, we have

$$\frac{\partial v_1}{\partial x_1} \approx \frac{1}{\Delta x \Delta y}((v_1)_L - (v_1)_R)\Delta y.$$

The generalization of this approach to unstructured grids leads to a rather complicated scheme, but there are two possible remedies. For cell-centered methods, the following heuristic approach is commonly used [44]. First, compute intermediate approximations of the gradients using arithmetics means:

$$\nabla \phi_i = \frac{1}{|\Omega_i|} \sum_{e_{ij} \in \partial \Omega_i} \frac{1}{2}(\phi_i + \phi_j)\mathbf{n}_{ij}|e_{ij}|.$$

Then, the means of these are taken:

$$\overline{\nabla \phi}_{ij} = \frac{1}{2}(\nabla \phi_i + \nabla \phi_j).$$

Finally, using the approximated directional derivative

$$\left(\frac{\partial \phi}{\partial l}\right)_{ij} := \frac{\phi_j - \phi_i}{|e_{ij}|}$$

and the normalized vector $\mathbf{r}_{ij}$, pointing from cell center i to cell center j, we obtain the final corrected approximation

$$\nabla \phi_{ij} = \overline{\nabla \phi}_{ij} - \left(\overline{\nabla \phi}_{ij} \cdot \mathbf{r}_{ij} - \left(\frac{\partial \phi}{\partial l}\right)_{ij}\right) \mathbf{r}_{ij}. \qquad (3.14)$$

Alternatively, if a structured cell-vertex method is employed, we can simply use (3.13) to obtain the derivative, whereas for dual cell Ω', we just use the original primal cell. For an unstructured cell-vertex method using triangles and tetrahedrons as primary cells, we can define a unique gradient on these based on the unique linear function interpolating the values in the nodes of the primary grid [198]. In the three-dimensional case we have four points $\mathbf{x}_1, ..., \mathbf{x}_4$ with the values there called $\phi_1, ..., \phi_4$. We then obtain for the components of the gradient

$$\nabla \phi_i = \frac{\det \mathbf{A}_i}{\det \mathbf{A}}, \qquad (3.15)$$

where

$$\mathbf{A} = \begin{pmatrix} (\mathbf{x}_4 - \mathbf{x}_1)^T \\ (\mathbf{x}_2 - \mathbf{x}_1)^T \\ (\mathbf{x}_3 - \mathbf{x}_1)^T \end{pmatrix}$$

and $\mathbf{A}_i$ is obtained by replacing the ith column of $\mathbf{A}$ with the vector $((\phi_2 - \phi_1), (\phi_3 - \phi_1), (\phi_4 - \phi_1))^T$. In the two-dimensional case, the formulas are accordingly derived from Cramer's rule.

3.4 Convergence theory for finite volume methods

3.4.1 Hyperbolic conservation laws

By contrast to finite element methods for elliptic equations, the convergence theory for numerical methods for conservation laws is unsatisfying, in particular for the relevant case of nonlinear systems in multiple dimensions. Nevertheless, during the last decade, significant advancements have been made, so that the theory for scalar one-dimensional equations reaches a certain maturity. An important property of finite volume schemes is that they are conservative, meaning that the total amount of a conservative quantity inside the computational domain is changed only through fluxes and source terms. Thus, the numerical scheme neither creates nor destroys or teleports mass. The theorem of Lax-Wendroff tells us that this is a good thing: If a scheme for a one-dimensional equation is conservative and consistent and the numerical solution is of bounded variation, then it converges to a weak solution of the conservation law, if it converges [172].

The total variation of a function in space is defined as

$$TV(u(x)) = \int |u'(x)|dx. \tag{3.16}$$

In the case of a grid function, this simplifies to

$$TV(u) = \sum_i |u_i - u_{i-1}|. \tag{3.17}$$

The space of functions in $\mathbb{R}^d$ with bounded total variation is called $BV(\mathbb{R}^d)$. Convergence proofs become possible using this concept, since a set of functions with a bounded total variation and a compact support is compact in L_1.

The Lax-Wendroff theorem can be improved to obtain convergence to the unique weak entropy solution (2.30) by requiring that the scheme satisfies a discrete version of the entropy inequality:

$$\eta(u_i^{n+1}) \leq \eta(u_i^n) - \frac{\Delta t}{\Delta x}(\Psi(u_j^n, u_{j+1}^n) - \Psi(u_{j-1}^n, u_j^n)). \tag{3.18}$$

Here, $\Psi(u, v)$ is a numerical entropy flux function that is consistent with the continuous entropy function ϕ in the sense of consistency of a numerical flux function.

Furthermore, a scheme is called monotone, if it has the following property: Given two numerical solutions u^n and v^n with $u^n \geq v^n$, it holds that $u^{n+1} \geq v^{n+1}$.

We then have the following theorem [164], which says that for a scalar equation, a monotone scheme converges to the weak entropy solution (2.30) of the equations and gives an a priori error estimate:

Theorem 3 *Let $u_0(x)$ be in $BV(\mathbb{R}^d)$ and $u(x,t)$ be a weak entropy solution of the conservation law. Furthermore, let u_h be a numerical solution obtained by a monotone scheme with explicit Euler integration in time with a suitably restricted time step. Let K be a specific subset of $\mathbb{R}^d \times \mathbb{R}^+$. Then there exists a constant $C > 0$, such that*

$$\|u - u_h\|_{L_1(K)} \leq Ch^{1/4}.$$

In the one-dimensional case, the statement goes with $h^{1/2}$ and this convergence rate is optimal.

The time step restriction in this theorem will be discussed further in the next chapter in section 4.2.

When we consider a one-dimensional linear system instead of a scalar equation, it is possible to apply the results to the diagonalized form of the equations and thus, they carry through. However, while diagonalization is still possible for nonlinear systems, it depends on the solution and thus, the equations do not truly decouple. Therefore, the convergence theory for nonlinear systems is very limited.

3.4.2   Parabolic conservation laws

In the case of parabolic conservation laws, there are even fewer results than for hyperbolic equations. This is counterintuitive, since a major problem for the analysis of hyperbolic conservation laws are the discontinuities, which are smoothed through viscous terms. However, the additional terms lead to other difficulties. For example, the total variation is not guaranteed to be nonincreasing for the exact solutions. There is extensive analysis of the linear convection diffusion equation. A more interesting result for nonlinear scalar conservation laws with a diffusion term in multiple dimensions is due to Ohlberger [214]. Since the result is very technical, we give only the gist of it. However, we will discuss the implied restriction on the time step for reasons of stability in section 4.2.

Theorem 4 *Consider the equation in $\mathbb{R}^d$*

$$u_t + \nabla \cdot f(u) = \nu \Delta u$$

with $1 \gg \nu > 0$ and reasonably regular initial data with compact support and a reasonably regular function f. Assume a finite volume discretization on a reasonable grid and a time step Δt that is small in a certain sense. Then there is a constant $K > 0$, such that

$$\|u - u_h\|_{L_1(\mathbb{R}^d \times \mathbb{R}^+)} \leq K(\nu + \sqrt{\Delta x} + \sqrt{\Delta t})^{1/2} \tag{3.19}$$

As in the case of hyperbolic equations, there are no results for the case of nonlinear systems in multiple dimensions, in particular not for the Navier-Stokes equations.

3.5 Source terms

Regarding source terms, there are two basic approaches. The first one will be discussed in the next chapter, and splits the source terms from the other terms to separately integrate these in time. This separation might not a good idea, if the fluxes and the source terms have to be in balance and it is the separation that prevents this balance from being achieved. Therefore, the other approach is to incorporate these into the computation of the fluxes. One example are the well-balanced schemes of Greenberg and Leroux [114] or the Z-wave method of Bale et al. [8].

3.6 Finite volume methods of higher order

The method as given so far uses a piecewise constant approximation to the solution $\mathbf{u}(\mathbf{x}, t)$ and therefore results in a method that can be at most of first order. This is not sufficient for practical applications, because the required spatial resolution renders the schemes inefficient. Therefore, more accurate methods are necessary that need a smaller amount of degrees of freedom at as little additional cost as possible. A large number of approaches to obtain higher order have been suggested. The standard approach is the MUSCL scheme. This uses a reconstruction technique to obtain a linear representation $\mathbf{u}_i(t)$ of $\mathbf{u}(\mathbf{x}, t)$ in each cell. Others are Weighted essentially non-oscillatory (WENO) and Arbitrary high order derivative Riemann problem (ADER) schemes, as well as DG methods, which will be described in section 3.7. First, we will explain important results on higher-order discretizations for conservation laws.

3.6.1 Convergence theory for higher-order finite volume schemes

It turns out that a monotone scheme can be at most of first order [123]. Therefore, this requirement is too strong and needs to be relaxed. This has led to the use of total variation diminishing (TVD) schemes, which means that the total variation of the numerical solution is nonincreasing with time, a property that is shared by the exact solutions in characteristic variables of hyperbolic equations. One point is that the TVD property implies that no new maxima and minima can be created, leading to nonoscillatory schemes.

It is possible to prove convergence results for a fixed time step Δt using the so-called TV-stability property, which means that the total variation of the numerical solution is bounded, independently of the grid. If a scheme is consistent, conservative, and TV-stable, then the solution converges in the

L_1-norm to a weak solution of the conservation law, which is not necessarily the entropy weak solution. Note that to prove whether a scheme has the TVD property, it is not sufficient to consider the space discretization alone, but the time discretization has to be considered as well. For the sake of simplicity, we will only consider explicit Euler time integration for the remainder of this chapter and examine different time integration methods in the next chapter.

Unfortunately, it turns out that even with this relaxed property, TVD schemes are not the final answer. First, Godunov proved that a linear TVD scheme can be at most of first order [108]. For this reason, higher-order schemes have to be nonlinear. Furthermore, they have to reduce to first order at spatial local maxima, as was proven by Osher and Chakravarty [216]. In multiple dimensions, as was proven by Goodman and LeVeque [109], TVD schemes are at most first order. Finally, on unstructured grids, not even a plane wave can be transported without increasing the total variation. Nevertheless, nonlinear TVD finite volume schemes using higher-order ideas have become the workhorse in academia and industry in the form of the MUSCL schemes using reconstruction and limiters.

An alternative development are positive schemes, as made popular by Spekreijse [259], respectively local extrema diminishing schemes (LED) as suggested by Jameson [143]. These were developments in parallel to the development of TVD schemes that looked particularly interesting in the context of unsteady flows. However, it turns out that these schemes have the same restrictions as TVD schemes. Nevertheless, this led to the development of interesting flux functions like CUSP and SLIP [143], which are inherently of higher order and widely used in aerodynamic codes.

3.6.2   Linear reconstruction

The idea of reconstruction is to use the given piecewise constant approximation at the given time level and to reconstruct a piecewise linear approximation from this with the same integral mean.

The procedure is based on the primitive variables $\mathbf{q} = (\rho, v_1, v_2, p)^T$, as this is numerically more stable than using the conservative variables [195]. At a given time t, the linear representation of a primitive variable $q \in \{\rho, v_1, ..., v_d, p\}$ in cell i with barycenter $\mathbf{x}_i$ is given by

$$q(\mathbf{x}) = q_i + \nabla q \cdot (\mathbf{x} - \mathbf{x}_i), \qquad (3.20)$$

where q_i is the primitive value corresponding to the conservative mean values in Ω_i. The unknown components of ∇q represent the slopes and are obtained by solving a least square problem. In the case of the Navier-Stokes equations, these slopes can be reused as gradients for the computation of the viscous fluxes.

For a cell-centered method on cartesian grids, the computation of the gradients is straightforward. If unstructured grids are employed, the following procedure is suitable. Let C be the closed polygonal curve that connects the

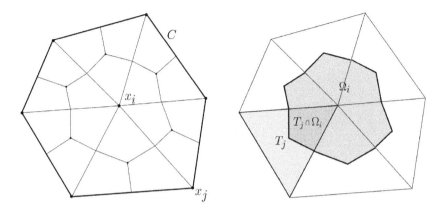

FIGURE 3.6: Left: The curve C used in the construction of the least squares problem for an unstructured cell. Right: The intersection of the primary triangle with a dual cell. (Credit: V. Straub.)

barycenters of the neighboring cells (see Figure 3.6 (left)). We then define the piecewise linear function q_c on C by setting

$$q_c(\mathbf{x}_j) = q_j$$

for all barycenters $\mathbf{x}_j$ of neighboring cells. The least squares problem, which has to be solved for all primitive variables, is then

$$\min_{\nabla q \in \mathbb{R}^d} L(\nabla q) := \int_C (q(\mathbf{x}) - q_c(\mathbf{x}))^2 ds. \tag{3.21}$$

For a cell-vertex scheme, first the unique linear interpolants of the cell averages located at the nodes are computed in each primary cell T_j [198], as described in (3.15) in the section on viscous terms. Let these have the gradient ∇q_j. The gradient on the dual box Ω_i is then defined as

$$\nabla q = \frac{1}{|\Omega_i|} \sum_j \nabla q_j |T_j \cap \Omega_i|. \tag{3.22}$$

Again, this procedure has to be done for each primitive variable. Note that $|T_j \cap \Omega_i|$ is zero except for the few primary cells that have a part in the creation of the box Ω_i (see Figure 3.6 (right)). Thus, the reconstruction procedure is significantly easier to implement and apply for cell-vertex methods than for cell-centered methods in the case of unstructured grids.

3.6.3 Limiters

As mentioned before, to obtain a TVD scheme, the scheme can be at most of first order at shocks and local extrema. Thus, on top of the reconstruction scheme, a so-called limiter function is needed. Typically a slope limiter ϕ is employed for the switching between first- and higher-order spatial discretization:

$$q(\mathbf{x}) = q_i + \phi \nabla q \cdot (\mathbf{x} - \mathbf{x}_i). \qquad (3.23)$$

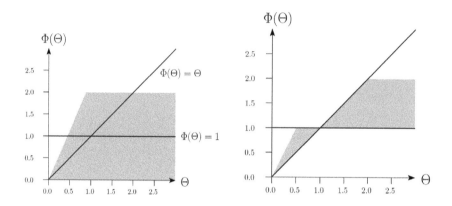

FIGURE 3.7: Left: Admissible region for a TVD slope limiter. Right: Admissible region for a second-order TVD slope limiter. (Credit: V. Straub.)

If the limiter function is zero, the discretization is reduced to first order, while it can be of second order for other values. In one dimension, conditions can be found for a limiter to result in a TVD scheme. For these, the limiter is taken as a function of the left and right slope and should be between 0 and 2, with the more precise shape given in Figure 3.7 (left). Furthermore, the conditions for the resulting scheme to be second order are demonstrated in Figure 3.7 (right). A large number of limiters has been suggested that fulfil these conditions, for example the superbee limiter or the van Albada limiter. On structured grids, these can be applied directionwise, and it is possible to prove the TVD property.

If the grid is unstructured, heuristic approaches must be used. We suggest either the Barth-Jesperson limiter [12] or the limiter proposed by Venkatakrishnan [291]. The latter is defined in cell i as follows:

$$\phi = \min_{j \in N(i)} \frac{(\Delta_{ij}^2 + \epsilon) + 2\Delta_{ij}\overline{\Delta}_{ij}}{\Delta_{ij}^2 + 2\overline{\Delta}_{ij}^2 + \Delta_{ij}\overline{\Delta}_{ij} + \epsilon}. \qquad (3.24)$$

Here, $\epsilon = 10^{-6}$ and

$$\Delta_{ij} = \begin{cases} 0, & \text{if } \overline{\Delta}_{ij}\Delta_{ij} < 0, \\ q_j - q_i, & \text{else,} \end{cases}$$

where the indicator $\overline{\Delta}_{ij}$ is given by

$$\overline{\Delta}_{ij} := q_{x_1}(x_{1_{ij}} - x_{1_i}) - q_{x_2}(x_{2_{ij}} - x_{2_i})$$

with the edge center $(x_{1_{ij}}, x_{2_{ij}})$.

Regarding boundaries, in the ghost cells at the inlet boundaries, the slopes and limiters are computed in the usual way. As the ghost cells are the definite border of the computational domain, no values beyond the ghost edges are interpolated or incorporated in any way, as is done for the fixed wall.

3.7   Discontinuous Galerkin methods

For a number of important problems, it is imperative to have a method of order greater than 2. An important case is the computation of turbulent flows using LES or DES. There, the resolution needed is so high that not only are high-order methods significantly more efficient than low order methods, but it is also unclear how a 3D LES could be obtained using a second-order scheme (remember that with a coarse resolution, the modeling errors of an LES make it worse than RANS). Another example would be trying to track vortices over a large time period. There, a scheme with high accuracy and little diffusion is necessary, if the number of grid points is to be kept reasonable. A problem of finite volume methods, in particular in 3D, is the current inability to obtain methods that are efficient and are of order higher than 2 and not extremely difficult to implement. A class of schemes that has received increasing research interest over the last 20 years are DG schemes [161, 152, 131], because there is reason to believe that this class can replace finite volume schemes for industrial applications where highly accurate computations are necessary.

DG schemes can be seen as more natural extensions of first-order finite volume schemes to higher orders than the reconstruction or WENO techniques mentioned earlier. Again, the solution is represented by a multivariate polynomial in each cell, leading to a cellwise continuous numerical solution. The specific polynomial is then determined using a Galerkin approach. There is a huge variety of methods based on this approach and a standard has only been established in parts. Furthermore, there are still a number of serious issues with DG methods that need to be solved before they are feasible for industrial applications [295]. In particular, there is the treatment of curved boundaries, the efficient solution of the appearing systems inside an implicit time integration, as well as the formulation of appropriate limiter functions.

Starting point of a DG method is the weak form of the conservation law, whereby as test functions, polynomials ϕ from some test space V_h are used:

$$\int_\Omega \mathbf{u}_t \phi d\Omega + \int_\Omega \nabla \cdot \mathbf{f}(\mathbf{u}, \nabla \mathbf{u}) \phi d\Omega = \mathbf{0}, \quad \forall \phi \in V_h.$$

We then use integration by parts to obtain

$$\int_\Omega \mathbf{u}_t \phi d\Omega + \int_{\partial\Omega} \mathbf{f} \cdot \mathbf{n}\phi ds - \int_\Omega \mathbf{f} \cdot \nabla\phi d\Omega = \mathbf{0}, \quad \forall \phi \in V_h. \tag{3.25}$$

At this point, the solution is approximated in every cell Ω_i by a polynomial

$$\mathbf{u}_i^P(\mathbf{x}, t) = \sum_j \mathbf{u}_{i,j}(t)\phi_j(\mathbf{x}), \tag{3.26}$$

where $\mathbf{u}_{i,j}(t) \in \mathbb{R}^{d+2}$ are coefficients and ϕ_j are polynomial basis functions in $\mathbb{R}^d$ of up to degree p. Thus, by contrast to a finite volume method, where we have just $d+2$ mean values $\mathbf{u}_i(t)$ as unknowns in cell Ω_i, there are now $d+2$ times the number of basis functions unknowns per cell for a DG method. We denote the dimension of this space by N.

Typically, the test functions are chosen from the same space. A specific DG method is obtained when we choose a polynomial basis and the quadrature rules for the integrals in the scalar products, in particular the nodes and weights. Generally, there are two different types of basis polynomials, namely modal and nodal bases. A nodal basis is defined by a number of nodes, through which then Lagrange polynomials are defined. On the other hand, a modal basis is defined by functions only, a prime example would be monomials. Typically, a modal basis is hierarchical and some authors use the term "hierarchical" instead.

Obviously, the polynomial basis depends on the shape of the cell. Therefore, it is common for DG methods to restrict the possible shapes to quadrangles, triangles, tetrahedrons, or cubes, then define the basis a priori for corresponding reference elements, for example, unit quadrangles, triangles, tetrahedrons, or cubes, to precompute as many terms as possible for that element. The basis for a specific cell is then obtained by a transformation from the reference cell. We will now demonstrate this for a curved quadrangle; the technique is similar for triangles.

First, the cell Ω_i is transformed to a unit cell, for example $[0, 1]^2$ with coordinates $\xi = (\xi_1, \xi_2) = (\xi, \eta)$ using an isoparametric transformation $\mathbf{r}$, which can be different for each cell. To this end, each of the four boundaries is represented by a polynomial $\Gamma_m(s)$, $m = 1, ..., 4$, $s \in [0, 1]$ and the four corners are denoted by $\mathbf{x}_j$, $j = 1, ..., 4$. The isoparametric transformation can then be understood as mapping the four corners and curves onto their representative in the reference space and filling the domain in between by convex combinations of the curves opposite of each other:

$$\begin{aligned}
\mathbf{r}_i(\xi, \eta) = &(1 - \xi)\Gamma_1(\xi) + \eta\Gamma_3(\xi) + (1 - \xi)\Gamma_4(\eta) + \xi\Gamma_2(\eta) \\
&- \mathbf{x}_1(1 - \xi)(1 - \eta) - \mathbf{x}_2\xi(1 - \eta) - \mathbf{x}_3\xi\eta - \mathbf{x}_4(1 - \xi)\eta. \quad (3.27)
\end{aligned}$$

For different reference cells, similar transformations can be employed, see [152, Chapter 4].

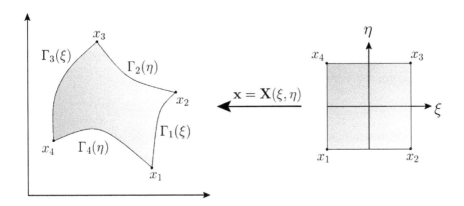

FIGURE 3.8: Illustration of the isoparametric mapping between an arbitrary quadrilateral cell and the reference element. (Credit: V. Straub.)

We thus obtain the new local equations

$$\mathbf{u}_t + \nabla_\xi \cdot \tilde{\mathbf{f}} = 0$$

with $\tilde{\mathbf{f}}_k = \frac{1}{J}(\mathbf{v}_l \times \mathbf{v}_m) \cdot \mathbf{f}$ (k, l, m to be cyclic). Here, $\mathbf{v}_i = \partial \mathbf{r}/\partial \xi_i$, but formally embedded into $\mathbb{R}^3$ for $i = 1, 2$ and $\mathbf{v}_3 = (0, 0, 1)^T$. Furthermore, $J = \mathbf{v}_k \cdot (\mathbf{v}_l \times \mathbf{v}_m)$ is the functional determinant. A derivation of this can be found in [161, Chapter 6]. Note that most of the above-mentioned geometric quantities can be precomputed and stored directly after grid generation.

We will now drop the tilde, but nevertheless now work on the reference cell Ω only. There, we define the quadrature rule to approximate the integrals in (3.25). Here, we choose Gaussian quadrature, such that polynomials of degree up to $2p$ are integrated exactly, which is well defined on the reference shapes with known nodes $\mathbf{x}_l$ and weights w_l:

$$\int_\Omega f(\mathbf{x}) d\Omega \approx \sum_{j=1}^N f(\mathbf{x}_j) w_j. \tag{3.28}$$

Applying these quadrature rules in (3.25), we obtain for the different terms:

$$\int_\Omega \mathbf{u}_t^P \phi_k d\Omega = \int_\Omega \frac{\partial}{\partial t} \sum_{j=1}^N \mathbf{u}_j(t) \phi_j(\mathbf{x}) \phi_k(\mathbf{x}) d\Omega, \quad k = 1, ..., N \tag{3.29}$$

which, due to the quadrature rule being exact for polynomials up to degree $2p$, can be written as $\mathbf{M}\bar{\mathbf{u}}_t$. Here, the mass matrix $\mathbf{M} \in \mathbb{R}^{(d+2)N \times (d+2)N}$ is a block diagonal matrix with blocks in $\mathbb{R}^{(d+2) \times (d+2)}$ where the value on the diagonal of the block (j, k) is constant and given by

$$M_{jk} = \int_\Omega \phi_j \phi_k d\Omega, \quad j, k = 1, ..., N.$$

Furthermore, the vector of coefficients is given by

$$\bar{\mathbf{u}} = (\mathbf{u}_1^T(t), ..., \mathbf{u}_N^T(t))^T \in \mathbb{R}^{(d+2)N}.$$

For the volume integral, we obtain using (3.28):

$$\int_\Omega \mathbf{f}(\mathbf{u}^P) \cdot \nabla \phi_k d\Omega = \int_\Omega \mathbf{f}\left(\sum_{j=1}^N \mathbf{u}_j(t)\phi_j(\mathbf{x})\right) \cdot \nabla \phi_k(\mathbf{x}) d\Omega \quad k = 1, ..., N$$

$$\approx \sum_{l=1}^N \mathbf{f}\left(\sum_{j=1}^N \mathbf{u}_j(t)\phi_j(\mathbf{x}_l)\right) \cdot \nabla \phi_k(\mathbf{x}_l)w_l \quad k = 1, ..., N.$$

$$(3.30)$$

This can again be written in compact form as $\sum_{j=1}^d \mathbf{S}_j \bar{\mathbf{f}}_j$ with $\bar{\mathbf{f}}_j$ being the vector of evaluations of the function $\mathbf{f}_j$ at the quadrature nodes and $\mathbf{S}_j \in \mathbb{R}^{(d+2)N \times (d+2)N}$ being again block diagonal matrices where on the diagonal of the block (k, l) we have the constant value

$$S_{j_{kl}} = \partial_{x_j} \phi_k(\mathbf{x}_l)w_l, \quad k, l = 1, ..., N.$$

Finally, the boundary integral can be written as

$$\int_{\partial\Omega} \mathbf{f} \cdot \mathbf{n}\phi_k dS \approx \sum_{Faces} \sum_{l=1}^{\hat{N}} \mathbf{f}\left(\sum_{j=1}^N \mathbf{u}_j(t)\phi_j(\mathbf{x}_l)\right) \cdot \mathbf{n}\phi_k(\mathbf{x}_l)w_l, \quad k = 1, ..., N,$$

$$(3.31)$$

with $\hat{N}$ being the number of quadrature points on the boundary. Here, the compact form is $\mathbf{M}^S \bar{\mathbf{g}}_j$ with $\mathbf{M}^S \in \mathbb{R}^{(d+2)N \times (d+2)N}$ block diagonal with values

$$M_{kl}^S = \phi_k(\mathbf{x}_l)w_l$$

on the diagonal of the block (k, l). The vector $\bar{\mathbf{g}}_j$ consists of function evalua-tions at the quadrature nodes on the surface. As in the case of finite volume schemes, this has to be done using a numerical flux function, thus coupling neighboring cells together. Since the spatial resolution is obtained primarily through the high-order polynomials, the specific flux function used is not as important as for finite volume schemes. For the inviscid fluxes, any of the previously discussed schemes can be used, whereas the choice of the viscous flux is slightly more complicated than before. This will be discussed in section 3.7.3.

Finally, we obtain the following ordinary differential equation for the co-efficients in a reference cell:

$$\mathbf{M}\bar{\mathbf{u}}_t + \sum_{j=1}^{nFaces} \mathbf{M}_j^S \bar{\mathbf{g}}_j - \sum_{j=1}^{d} \mathbf{S}_j \bar{\mathbf{f}}_j = \mathbf{0}. \tag{3.32}$$

Note that the matrices $\mathbf{M}$, $\mathbf{M}^S$, and $\mathbf{S}_j$ depend only on the basis and the geometry, not on $\mathbf{u}$, and can thus be precomputed after grid generation.

Obviously, the choice of basis is very important for the question of effi-ciency of the resulting methods. Essentially, it turns out that for an efficient implementation of DG schemes, a nodal basis is more useful. This is because due to the use of Lagrange polynomials, a nodal basis allows the computation of the vectors $\bar{\mathbf{f}}_j$ and $\bar{\mathbf{g}}_j$ without the need for evaluating the ansatz function (3.26) at those points. In a similar fashion, some other approaches to obtain higher-order methods also lead to a scheme that is similar to the nodal DG method, in particular the flux reconstruction schemes of Huynh [139] and the energy-stable subfamily of schemes found by Vincent et al. [294]. Here, we con-sider two particular types of DG schemes, namely the DG spectral element method (DG-SEM) approach of Kopriva [162] and the modal-nodal method suggested by Gassner et al. [102].

Independently of the specific DG method chosen, the implementation has an important difference to that of a finite volume method, coming from the volume integral represented by the last term in (3.32). This is local to a cell and thus is treated by a loop of all cells. This part of the method has high locality and is embarrassingly parallel, contributing to DG methods being very suitable for high performance computing. Regarding the loop of the edges, the implementation is the same as for finite volume methods.

3.7.1 Polymorphic modal-nodal scheme

The modal-nodal approach allows for arbitrary element types and starts with a modal basis of monomials that are based on the barycenters of refer-ence cells. These are then orthonormalized using a Gram-Schmidt procedure. Thus, the basis is orthonormal and hierarchical, meaning that the basis func-tions can be sorted by the polynomial order. The dimension of the polynomial space is given by $N = (p+d)!/(p!d!)$, which amounts to 280 degrees of freedom in 3D for fifth-order polynomials. Generally, this basis can be used to define a DG scheme as is. In this case, the vectors $\bar{\mathbf{f}}_j$ and $\bar{\mathbf{g}}_j$ contain a huge num-ber of evaluations of the polynomial approximations (3.26). Therefore, this is combined with a nodal quadrature that reduces this computational effort significantly.

Thus, in addition, a nodal basis is defined for the reference elements chosen, where we use Legendre-Gauss-Lobatto (LGL) points on the edges. For the interior points, an LGL-type approach is suggested in [102]. Other choices are possible, but the LGL-type approach leads to a very small condition number of the basis. The modal coefficients $\hat{u}$ and the nodal coefficients $\tilde{u}$ can then be related using a Vandermonde matrix $\mathbf{V}$, which contains the evaluations of the modal basis polynomials at the nodes defining the nodal basis:

$$\tilde{u} = \mathbf{V}\hat{u}. \tag{3.33}$$

If the number of basis functions is not identical, as is the case for triangles and tetrahedrons, this matrix is rectangular. Then, the backtransformation $\mathbf{V}^{-1}$ is defined in a least squares sense, as the pseudoinverse. Note that in both cases, these operators depend on the geometry and choice of basis alone, and thus can be precomputed and stored. All in all, we obtain the following ordinary differential equation for a reference element:

$$\hat{u}_t = -\mathbf{V}^{-1}\tilde{\mathbf{M}}^{-1}\left(\sum_{i=1}^{nFaces}\tilde{\mathbf{M}}_i^S\tilde{g}_i - \sum_{k=1}^{d}\tilde{\mathbf{S}}_k\tilde{f}_k\right), \tag{3.34}$$

where the $\sim$ denotes terms based on the nodal basis.

3.7.2　DG Spectral Element Method

The DG-SEM requires quadrilateral and hexahedral cells, respectively, in three dimensions. However, hanging nodes as well as curved boundaries are possible. The method uses a nodal basis defined by Gaussian quadrature nodes. Thus, the dimension of the polynomial space is $N = (p+1)^d$. For the rest of the derivation, we will now assume a two-dimensional problem, which simplifies the formulas significantly. The extension to three dimensions is straightforward. On the unit rectangle $\Omega_u = [-1,1]$, the approximation is then written in local coordinates (ξ, η) as

$$\mathbf{u}(\xi,\eta) \approx \mathbf{u}^P(\xi,\eta,t) = \sum_{\mu=0}^{p}\sum_{\nu=0}^{p}\mathbf{u}_{\mu\nu}(t)\phi_{\mu\nu}(\xi,\eta), \tag{3.35}$$

with

$$\phi_{ij}(\xi,\eta) = l_i(\xi)l_j(\eta)$$

and Lagrange polynomials

$$l_j(\xi) = \Pi_{i=0,i\neq j}^{p}\frac{(\xi-\xi_i)}{\xi_j - \xi_i}$$

based on the nodes ξ_j, η_j of a Gauss quadrature. Note that we thus have to compute and store the metric terms from above per Gauss point per cell. Additionally, we use the interpolating ansatz

$$\mathbf{f}^P(\xi,\eta,t) = \sum_{\mu,\nu=0}^{p}\mathbf{f}_{\mu\nu}(t)\phi_{\mu\nu}(\xi,\eta) \tag{3.36}$$

with

$$\mathbf{f}_{ij}(t) = \mathbf{f}(\mathbf{u}_{ij}(t), \nabla \mathbf{u}_{ij}(t)).$$

Regarding the viscous terms, we need to compute gradients of the solution as well. We have for the gradient of the jth component:

$$\nabla u_j^P(t, \xi, \eta) = \left(\begin{array}{c} \sum_{\mu,\nu=0}^{p} u_{\mu\nu_j}(t) l_\mu'(\xi) l_\nu(\eta) \\ \sum_{\mu,\nu=0}^{p} u_{\mu\nu_j}(t) l_\mu(\xi) l_\nu'(\eta) \end{array} \right).$$

The scalar products in (3.25) are approximated using Gauss quadrature with the weights w_i, w_j:

$$\int_{-1}^{1} \int_{-1}^{1} v(\xi, \eta) d\xi d\eta \approx \sum_{i,j=0}^{p} v(\xi_i, \eta_j) w_i w_j.$$

Thus, the final approximation in one cell is given by

$$\int_{\Omega_u} \mathbf{u}_t^P \phi_{ij} d\Omega + \int_{\partial \Omega_u} \mathbf{f}^P \cdot \mathbf{n} \phi_{ij} ds - \int_{\Omega_u} \mathbf{f}^P \cdot \nabla \phi_{ij} d\Omega = \mathbf{0}, \quad i, j = 0, ..., p.$$

Due to the choice of using the Lagrange polynomials based on the Gauss-Legendre points, the terms simplify significantly. We obtain

$$\int_{\Omega_u} \mathbf{u}_t^P \phi_{ij} d\Omega = \frac{d\mathbf{u}_{ij}}{dt} w_i w_j$$

and

$$\int_{\Omega_u} \mathbf{f}^P \cdot \nabla \phi_{ij} d\Omega = \sum_\mu \mathbf{f}_{1_{\mu j}} l_i'(\xi_\mu) w_\mu w_j + \sum_\nu \mathbf{f}_{2_{i\nu}} l_j'(\eta_\nu) w_i w_\nu.$$

Regarding the flux, the boundary integral is the sum of the four integrals on the edges of the unit cell. To illustrate, we have for the lower edge:

$$\int_{-1}^{1} \mathbf{f}^P(\xi, -1) \cdot \mathbf{n} \phi_{ij}(\xi, -1) d\xi = -l_j(-1) \sum_{\mu,\nu=0}^{p} \mathbf{f}_{2_{\mu\nu}} l_\nu(-1) \int_{-1}^{1} l_\mu(\xi) l_i(\xi) d\xi.$$

Approximating this using the quadrature formula on the line, we obtain

$$\int_{-1}^{1} \mathbf{f}^P(\xi, -1) \cdot \mathbf{n} \phi_{ij}(\xi, -1) dx \approx -l_j(-1) \sum_{\mu,\nu=0}^{p} \mathbf{f}_{2_{\mu\nu}} l_\nu(-1) \sum_{a=0}^{p} l_\mu(\xi_a) l_i(\xi_a) w_i$$

$$= -\mathbf{f}_2^P(\xi_i, -1) l_j(-1) w_i.$$

Dividing by $w_i w_j$, we finally obtain

$$\frac{d\mathbf{u}_{ij}}{dt} + \left[\mathbf{f}_1^P(1, \eta_j) \frac{l_i(1)}{w_i} - \mathbf{f}_1^P(-1, \eta_j) \frac{l_i(-1)}{w_i} - \sum_\mu \mathbf{f}_{1_{\mu j}} \frac{l_i'(\xi_\mu) w_\mu}{w_i} \right]$$

$$+ \left[\mathbf{f}_2^P(\xi_i, 1) \frac{l_j(1)}{w_j} - \mathbf{f}_2^P(\xi_i, -1) \frac{l_j(-1)}{w_j} - \sum_\mu \mathbf{f}_{2_{i\mu}} \frac{l_j'(\eta_\mu) w_\mu}{w_j} \right] = 0 \quad (3.37)$$

or in three dimensions:

$$\frac{d\mathbf{u}_{ijk}}{dt} + \left[\mathbf{f}_1^P(1, \eta_j, \zeta_k) \frac{l_i(1)}{w_i} - \mathbf{f}_1^P(-1, \eta_j, \zeta_k) \frac{l_i(-1)}{w_i} - \sum_\mu \mathbf{f}_{1_{\mu jk}} \frac{l_i'(\xi_\mu) w_\mu}{w_i} \right]$$

$$+ \left[\mathbf{f}_2^P(\xi_i, 1, \zeta_k) \frac{l_j(1)}{w_j} - \mathbf{f}_2^P(\xi_i, -1, \zeta_k) \frac{l_j(-1)}{w_j} - \sum_\mu \mathbf{f}_{2_{i\mu k}} \frac{l_j'(\eta_\mu) w_\mu}{w_j} \right]$$

(3.38)

$$+ \left[\mathbf{f}_3^P(\xi_i, \eta_j, 1) \frac{l_k(1)}{w_k} - \mathbf{f}_3^P(\xi_i, \eta_j, -1) \frac{l_k(-1)}{w_k} - \sum_\mu \mathbf{f}_{3_{ij\mu}} \frac{l_k'(\zeta_\mu) w_\mu}{w_k} \right] = \mathbf{0},$$

The sums in (3.37) and (3.38) can be computed using information given in that cell. However, the boundary term needs to be coupled with the neighbors and since the numerical solution is discontinuous at the boundary by construction, again numerical flux functions are needed. We thus replace the boundary terms via

$$\mathbf{f}_1^P(1, \eta_j, \zeta_k) \approx \mathbf{f}^N(\mathbf{u}^P(1, \eta_j, \zeta_k); \hat{\mathbf{u}}^P(-1, \eta_j, \zeta_k); \mathbf{n}),$$

where $\hat{\mathbf{u}}^P$ corresponds to the polynomial from the neighboring cell.

3.7.3   Discretization of the viscous fluxes

More difficulty is posed by the diffusive terms, since the averaging procedure used in the finite volume case is unstable in this context. Several options to circumvent this problem are available, for example the Bassi-Rebay-flux in versions 1 and 2 [15], local DG of Cockburn and Shu [61], the compact DG flux (CDG) of Persson and Perraire [223], and CDG2 of Brdar et al. [47]. We will use the diffusive Generalized Riemann Problem (dGRP) flux of Gassner et al. [99, 100], which, as CDG, has the advantage of using a small discretization stencil. To derive this flux, we first have to go back to the volume integral in (3.25) and rewrite the viscous part (2.18), using the notation $\mathbf{f}^v = \mathbf{D}(\mathbf{u})\nabla\mathbf{u}$ and a second integration by parts, as

$$\int_\Omega \mathbf{D}(\mathbf{u})\nabla\mathbf{u} \cdot \nabla\phi \, d\Omega = \int_{\partial\Omega} \mathbf{u} \cdot ((\mathbf{D}(\mathbf{u})^T \nabla\phi) \cdot \mathbf{n}) ds - \int_\Omega \mathbf{u} \cdot (\nabla \cdot \mathbf{D}(\mathbf{u})^T \nabla\phi)) d\Omega.$$

(3.39)

We now introduce the adjoint flux

$$\mathbf{f}^{vv}(\mathbf{u}, \nabla\phi) := \mathbf{D}^T(\mathbf{u})\nabla\phi, \qquad (3.40)$$

which is later used to make sure that the discretization is consistent with the adjoint problem. A failure to do so results in either nonoptimal convergence rates or higher condition numbers [4]. To obtain a more efficient evaluation of the second term in (3.39), we use a third integration by parts

$$\int_{\Omega} \mathbf{u} \cdot (\nabla \cdot \mathbf{D}(\mathbf{u})^T \nabla \phi)) d\Omega = \int_{\partial \Omega} [\mathbf{u} \cdot (\mathbf{n} \cdot \mathbf{f}^{vv}(\mathbf{u}, \nabla \phi))]_{INT} ds - (\mathbf{f}^v, \nabla \phi), \quad (3.41)$$

where the subscript INT refers to an interior evaluation of the respective function at the boundary. Combining (3.39) and (3.41) in (3.25) we obtain

$$(\mathbf{u}_t, \phi) + \int_{\partial \Omega} \mathbf{f} \cdot \mathbf{n} \phi ds - (\mathbf{f}, \nabla \phi) + \int_{\partial \Omega} \mathbf{h}(\mathbf{u}, \mathbf{n}, \nabla \phi) ds = 0 \quad (3.42)$$

with the additional diffusion flux

$$\mathbf{h}(\mathbf{u}, \mathbf{n}, \nabla \phi) := \mathbf{u} \cdot (\mathbf{n} \cdot \mathbf{f}^{vv}(\mathbf{u}, \nabla \phi)) - [\mathbf{u} \cdot (\mathbf{n} \cdot \mathbf{f}^{vv}(\mathbf{u}, \nabla \phi))]_{INT}. \quad (3.43)$$

We can now formulate the dGRP flux by defining the numerical fluxes used in the computation of the boundary integrals in (3.42). For the viscous fluxes, this is derived from the viscous Riemann problem. If we assume that the flux has been rotated into normal direction (which is possible since the adjoint flux is rotationally invariant as well) and using the characteristic length scale

$$\Delta x_{ij} = \frac{\min(|\Omega_i|, |\Omega_j|)}{|\partial \Omega_{ij}|},$$

a constant η and the averaging operator

$$\{\mathbf{u}\} = 0.5(\mathbf{u}_L + \mathbf{u}_R), \quad (3.44)$$

the dGRP flux for the viscous term is given by

$$\mathbf{f}^{dGRP}(\mathbf{u}_L, \mathbf{u}_R) = \mathbf{f}_1^v \left(\{\mathbf{u}\}, \left(\frac{\eta}{\Delta x_{ij}} (\mathbf{u}_L - \mathbf{u}_R) + \{\mathbf{u}_n\}, \{\mathbf{u}_t\} \right)^T \right), \quad (3.45)$$

where $\mathbf{u}_t$ are the tangential components of $\mathbf{u}$. The constant η can be considered

p	1	2	3	4	5
β^*	1.5	0.7	0.35	0.25	0.13

TABLE 3.1: Parameter β^* in the dGRP scheme.

as a penalization of the jumps and is given by

$$\eta = \frac{2p + 1}{\sqrt{\pi} \beta^*(p)}$$

with β^* a real number which becomes smaller with increasing order p, see Table 3.1. Furthermore, for the adjoint flux (3.43), the approximation

$$\mathbf{h} = \frac{1}{2}(\mathbf{u}_L - \mathbf{u}_R) \cdot (\mathbf{f}^{vv}(\{\mathbf{u}\}, \nabla \Phi_L) \cdot \mathbf{n}) \quad (3.46)$$

is used.

3.8   Convergence theory for DG methods

In the case of DG methods, the theory does not use the total variation stability to prove convergence, as is the case for finite volume methods. Instead, energy stability is employed. This is natural, since the method uses a variational formulation in the first place. This way, convergence of higher order can be proved for a wide range of equations, see e.g. [131]. In particular, for scalar equations on cartesian and particular structured triangular grids, the optimal convergence order of $p + 1$ can be proved, whereas on unstructured grids, at most $p + 1/2$ is obtained.

Zhang and Shu could prove using the energy method, that for smooth solutions of a symmetrizable hyperbolic system in multiple dimensions, a DG scheme on triangles using an appropriate (SSP, as explained later) time integration scheme and an appropriate flux, converges in L_2 with a spatial order of $p + 1/2$ [305].

As for finite volume schemes, these results are only valid under a time step restriction, which turns out to be more severe and to depend on the polynomial degree. This will be discussed in section 4.2.

3.9   Boundary conditions

If an edge is part of the boundary of the computational domain, the numerical flux cannot be defined as before, since there is no state on the other side of the edge. There are several ways of handling this. Firstly, we can define a solution value; secondly, we can prescribe a flux; both on the edge. An alternative is the perfectly matched layer technique (PML) [2]. In a cell-vertex method, unknowns are located directly at the boundary. This means that when prescribing values there, these unknowns could be removed from the scheme, whereas when prescribing a flux, the unknowns remain. In the latter case, a situation can arise where prescribed flux corresponds to the flux of a value determined by a boundary condition, but the numerical solution itself does not satisfy that condition. This might sound strange at first, but since the numerical solution is an approximation in the first place, it is not clear that enforcing the boundary conditions strictly leads to an overall better numerical solution. In fact, for schemes that enforce the boundary condition, only weakly stability in L_2 can be obtained. This approach is natural for schemes where an unknown is placed directly at the boundary, such as cell-vertex finite volume schemes or some DG methods.

3.9.1 Implementation

Regarding implementation, there are again two possibilities. The first one is to add a layer of ghost cells outside of the computational domain. In these, values are prescribed before doing any flux computations. The boundary edges are just included in the set of all edges that one loops over. The value in the ghost cell is chosen, such that when the numerical flux is computed, a boundary condition is realized. The ghost cell approach is rather straightforward for cell-centered schemes, even for unstructured grids. For cell-vertex methods, it is rather unwieldy, since the boundary edges are special cases during grid generation. Adding ghost cells makes this even more complicated.

The alternative is to have each edge have a type, either *inner* or a boundary type. During the loop over all edges, the type is checked and the flux computation is chosen accordingly. For an inner edge, the flux computation remains as is, but for all other edge types, the computation of the flux is adjusted. For cell-vertex schemes, this is best realized by using a binary description of the boundary type on the primary grid. To this end, store the boundary type for each vertex of the primary grid. A boundary edge of the dual grid then has the boundary type of the AND of the boundary types of the connecting points.

3.9.2 Stability and the SBP-SAT technique

On an unbounded domain, it is not clear that (semi)-discrete solutions remain bounded with time, in particular not for the L_2 norm. On a bounded domain however, we intuitively expect this, and it is clear that the boundary conditions play a big role in this aspect. A theory that has been rather successful in identifying good boundary conditions is the SBP-SAT strategy (Summation by parts-Simultaneous Approximation Term) [265, 266]. The idea is to derive boundary conditions by requiring stability in the L_2 norm for linearized problems, using the energy method.

For a linear problem, a bounded solution implies to have a nonpositive time derivative in a norm. To illustrate the technique, we now assume a one-dimensional scalar problem with the unknowns ordered from left to right. A summation-by-parts (SBP) operator is one that discretely reflects integration by parts. More precisely, it is a discretization that can be written as

$$\mathbf{u}_t + \mathbf{P}^{-1}\mathbf{Q}\mathbf{u} = \mathbf{P}^{-1}\mathbf{S},$$

where $\mathbf{Q} + \mathbf{Q}^T = \mathbf{B} = \text{diag}(-1, 0, ..., 0, 1)$ gives a difference of boundary values, $\mathbf{P}$ is symmetric and positive definite and thus defines a norm, and $\mathbf{s}$ is the simultaneous approximation term (SAT) containing the boundary treatment. It has been shown that a lot of discretizations satisfy this splitting, in particular high-order finite difference schemes [53], cell-vertex finite volume methods [212], and DG-SEM [101], thus making the technique relevant.

After multiplying from the left with $\mathbf{P}$ and adding the transpose of the whole equation, we obtain an estimate of the $\mathbf{P}$-norm:

$$\frac{d}{dt}\|\mathbf{u}\|_P = \mathbf{u}_t\mathbf{Pu} + \mathbf{uPu}_t = -\mathbf{u}^T(\mathbf{Q}+\mathbf{Q}^T)\mathbf{u} + 2\mathbf{u}^T\mathbf{s}.$$
$$= u_m^2 - u_1^2 + 2\mathbf{u}^T\mathbf{s}$$

To make this term overall negative, the SAT $\mathbf{s}$ has to and can be chosen appropriately. Otherwise put, if the discretization corresponds to an SBP operator, it is a stable discretization in the interior and problems can arise only from the treatment of the boundary. The SAT is given by

$$\mathbf{s} = \sigma(\mathbf{u}-\mathbf{g}),$$

where $\sigma \in \mathbb{R}$ is a penalty parameter and $\mathbf{g}$ is the prescribed boundary condition. It is of note that enforcing $\mathbf{u} = \mathbf{g}$ does not lead to stability. Thus, the SBP-SAT technique uses a weak enforcement of the boundary condition. Boundary conditions that lead to stability have been derived for many linearized equations. Their application within a nonlinear problem is straightforward and while the stability result no longer applies, they have been proven to work well in practice.

3.9.3   Fixed wall

At a fixed wall in the case of the Euler equations, we use slip conditions as in section 2.7 for the continuous equations. Thus, there should be no flow through the boundary, but tangential to the boundary, no conditions are imposed. Therefore, at the evaluation points on the wall the condition $\mathbf{v}\cdot\mathbf{n} = 0$ has to hold. For the Navier-Stokes equations, we have to use no-slip boundary conditions; thus, we require the solution to have zero velocity in the boundary points: $\mathbf{v} = \mathbf{0}$. In addition, a boundary condition for the heat flux term has to be given. This is again done corresponding to the continuous equations as either isothermal or adiabatic, thus either prescribing a value for temperature or for the heat flux.

For the slip conditions using ghost cells, we have to prescribe a negative normal velocity, such that this adds up to zero on the wall. All other values are chosen the same as in the neighboring cell. Using boundary fluxes, the prescribed flux is $\mathbf{f} = (0, n_1 p, ..., n_d p, 0)^T$, where the pressure is a possibly extrapolated value at the boundary. Regarding no-slip conditions, the convective boundary flux is the same and there is no additional contribution from the viscous stress tensor, since the velocity is zero. In a ghost cell implementation, this would be realized by setting all velocity components to the appropriate negative value. Note that the slip condition can be used in the case of symmetric flows to cut the computational domain in half. To this end, the symmetry axis has to be chosen as a boundary of the domain.

In the case of an adiabatic wall condition, the additional flux term on top of the no-slip flux would be zero and similarly, nothing needed to be changed about the ghost cells. For an isothermal wall, the temperature values are extrapolated in the ghost cell implementation, whereas for boundary fluxes, a corresponding temperature gradient is determined.

These fluxes both satisfy the SBP-SAT framework and are entropy-stable for both the Euler and Navier-Stokes equations [267, 220].

3.9.4   Inflow and outflow boundaries

Other boundaries are artificial ones where the computational domain ends, but not the physical one. Thus, a boundary flux has to be found that leads to the correct solution in the limit of mesh width to zero, or that otherwise spoken, makes the artificial boundary interfere as little as possible with the solution on the larger domain.

Several types of boundary conditions are possible and should be used depending on the problem to solve. The most simple ones are *constant interpolation boundary conditions*, which means that we use Neumann boundary conditions where we set the derivative on the edge to zero. Using ghost cells, this corresponds to copying the value from the interior cell to the ghost cell. However, this reduces the order at the boundary to 1 and leads to problems when significant tangential flows are present.

Another option are *far field boundary conditions*, where the value in the ghost cell is always set to the initial value. This type of boundary conditions is often used for the computation of steady flows. However, it can happen that in this way, waves are reflected back into the computational domain. Therefore, it must be made sure that the boundary is sufficiently far away from regions where something happens, even for the computation of steady states. This makes the use of these conditions limited for unsteady flows.

To circumvent this problem, *nonreflecting boundary conditions* can be used [273] (also called *absorbing boundary conditions*). Essentially, we want outgoing waves to leave the computational domain without any disturbances reflecting backwards. This can be achieved by setting all incoming waves to 0.

Another possibility are *characteristic boundary conditions*, where depending on whether we are at an inlet or outlet boundary, we prescribe three, respectively one value in the ghost cell neighboring the boundary and the values from the computational domain are used for the remaining variables. For the Navier-Stokes equations, we have to prescribe all values at an inflow boundary, but 1 less at an outflow boundary, mentioned in section 2.7.

However, even for the Navier-Stokes equations, often the conditions for the Euler equations are employed. Although it cannot be proven that this leads to a well-posed problem, it works quite well. The intuitive explanation is that away from turbulent structures and boundary layers, the second-order terms can be neglected and thus, the Euler equations and their boundary conditions provide a very good approximation. We will now explain these in more detail.

As mentioned in section 2.7, the number of conditions to pose depends on the Mach number, respectively on the sign of the eigenvalues (2.27). Thus, for supersonic inflow, the farfield value determines the flux on the boundary, whereas for supersonic outflow, the value in the computational domain defines the boundary flux. For subsonic flow, conditions can be derived using the theory of characteristics. A linear approach using the SBP-SAT theory can be found in [212]. Here, we describe a nonlinear variant employing Riemann invariants [301]. Given a value $\mathbf{u}_i$ in the cell next to the boundary, the value $\mathbf{u}_b$ in the ghost cell on the other side of the boundary with normal vector $\mathbf{n}$ has to be defined. The conditions obtained in this way are not unique, but depend on the components of the farfield values $\mathbf{u}_0$ used. At an inflow boundary, we obtain for the case that ρ_b and $\mathbf{v}_b$ are prescribed

$$
\begin{aligned}
\rho_b &= \rho_b, \\
\mathbf{m}_b &= \rho_b \mathbf{v}_b, \\
\rho E_b &= \rho_b((v_{i_n} + 2c_i/(\gamma - 1) - v_{b_n})/(2\gamma) + |\mathbf{v}_b|^2/2).
\end{aligned}
$$

For subsonic outflow in the case that the pressure p_b is prescribed, we have, using the entropy (2.29)

$$
\begin{aligned}
\rho_b &= \left(p_b e^{-s_i/c_v}\right)^{1/(\gamma-1)} \\
\mathbf{m}_b &= \rho_b(\mathbf{v}_i + 2c_i/(\gamma - 1)\mathbf{n}) \\
\rho E_b &= \rho_b(p_i/((\gamma - 1)\rho_b) + |\mathbf{m}_b/\rho_b|^2/2).
\end{aligned}
$$

Alternatively, one can prescribe the flux obtained when evaluating it in $\mathbf{u}_b$.

3.9.5   Periodic boundaries

Another possibility are periodic boundary conditions (2.32). These are an easy way to obtain long time numerical test problems that do not require a large computational domain. When using ghost cells, periodic boundary conditions are implemented by copying the value from the cells on the one periodic side to the ghost cells of the other periodic side and vice versa. With fluxes, these are computed based on the values in the cells on both sides of the periodic boundary. When using a cell-vertex method with a dual grid, things become slightly more complicated. One way is to define cells based on the vertices on the boundary, that then have volume on both sides. In this way, no fluxes along the periodic boundary need to be computed.

3.10   Spatial adaptation

Though important, the topic of spatial adaptation is not a focus of this book. We will nevertheless give a few pointers to the important concepts in this field.

As mentioned before, the convergence theory for nonlinear systems of conservation laws for both finite volume and DG methods is lacking and therefore, rigorous error estimates similar to theorem 3 do not exist for the interesting methods and equations. In fact, it is known for the Euler equations that nonunique solutions may exist and for the Navier-Stokes equations, uniqueness has not been proven for general initial conditions. Therefore, by contrast to finite element methods for elliptic PDEs, grid adaptation has to be based on more or less heuristical error estimators.

For steady state computations, a large amount of literature is available on grid adaptation. There, it is clear that any error in the solution is due to the space discretization. Therefore, grid adaptation can be done by starting on a coarse grid, computing the steady state on that grid, estimating the error of that solution, refining the mesh, and iterating these steps.

In the unsteady case, we have both spatial and temporal discretization errors. In the context of the method of lines, it is natural to try to estimate and treat these separately. We will explain in the next chapter how to get cheap estimates of the time discretization error for a given ordinary differential equation defined by the spatial discretization. The spatial error would be estimated separately and a grid refinement performed every couple of time steps. An alternative are space-time discretization methods that will be briefly discussed in section 4.9.

As error estimators for steady state problems, the residual can be used, although it is known that this can be a misleading indicator for the local error for nonlinear problems. Then, there is the popular method of gradient-based grid adaptation. There, the assumption is made that large gradients correspond to large local discretization errors, which is valid for smooth solutions, e.g., subsonic flows. For transonic or supersonic flows which typically contain shocks, gradients can even increase with grid adaptation, leading to a better accuracy, but a worse error estimation. Nevertheless, this leads to reasonably good results in practice. Gradient-based adaptation can also be used for unsteady flow problems.

Another important idea for adaptivity are goal-oriented methods that adapt the grid to decrease errors in a user-specified, typically nonlinear functional $J(\mathbf{u})$. The functional reflects a quantity that the user is particularly interested in, for example lift over drag, but it can also be the error. Often, the solution of the PDE is of minor interest and only the means to compute the value of a functional. The most widely used technique is the dual weighted residual approach (DWR) [16]. There, the error in the value of the functional

$J(\mathbf{u}^*) - J(\mathbf{u}_h)$ is estimated, where $\mathbf{u}^*$ is the solution and $\mathbf{u}_h$ a discrete approximation. This is done by solving a linearized form of the so-called dual or adjoint problem. For time-adaptive problems, this is a linear time-dependent problem backwards in time. This makes the grid adaptation process rather costly, but if successful, leads to a powerful refinement strategy [124, 70]. Another question is to find a functional other than the residual that leads to an overall good solution to the complete problem. To this end, Fidkowski and Roe suggest to use an entropy functional [91].

Finally, a strategy for the actual grid refinement has to be chosen. A famous approach is due to Berger and Collela and is known as AMR (for Adaptive Mesh Refinement) [20]. The idea is to work with cartesian structured meshes and to have local, rectangular cartesian patches, where the mesh is refined. These patches can again have refined sub patches and so on. This has been implemented in the software ClawPack [183]. When using high order methods, an alternative is called p-refinement, to distinguish from h-refinement, where the mesh is refined. In p-refinement, the order is varied instead, thus locally steering resolution [5]. Combining both approaches is known as hp-adaptivity. Not much work is available on this topic for DG methods [71, 1].

Chapter 4

Time integration schemes

After the discretization in space, when combining the equations for one cell in one coupled system, we obtain a large system of ordinary differential equations (ODEs) or more precisely, an initial value problem (IVP)

$$\frac{d}{dt}\underline{u}(t) = \underline{f}(t, \underline{u}(t)), \quad \underline{u}(t_0) = \underline{u}^0, \quad t \in [t_0, t_{end}], \tag{4.1}$$

where $\underline{u} \in \mathbb{R}^m$ is a vector containing all the unknowns from all grid cells. Very often in CFD, the function $\underline{f} : \mathbb{R}^{m+1} \to \mathbb{R}^m$ defined by the spatial discretization has no explicit dependence on t and then we obtain an autonomous IVP:

$$\frac{d}{dt}\underline{u}(t) = \underline{f}(\underline{u}(t)), \quad \underline{u}(0) = \underline{u}^0, \quad t \in [t_0, t_{end}]. \tag{4.2}$$

There is a huge variety of numerical methods to solve problems of this type; see for example the classic textbooks [121, 122]. Typically, these provide a number of approximations $\underline{u}^n \approx \underline{u}(t_n)$ at discrete times $t_0, t_1, ..., t_{end}$. Using the generating function Φ of the numerical method, they can be written as

$$\underline{u}^{n+1} = \underline{u}^n + \Delta t \Phi, \quad t_{n+1} = t_n + \Delta t.$$

If Φ is a function of data points at past times, it is called a multistep method, otherwise it is a onestep method. Multistep methods need some sort of starting procedure, for example a onestep method. Furthermore, a scheme is called implicit if it incorporates the unknown data $\underline{u}^{n+1}$, otherwise it is called explicit. Implicit schemes require solving an equation system in every step. In this chapter, we will assume that these can be solved to whatever accuracy is needed. Methods for doing so, as well as the problems appearing there, are discussed in the next chapter.

The prototypical schemes are the explicit Euler method

$$\underline{u}^{n+1} = \underline{u}^n + \Delta t \underline{f}(\underline{u}^n) \tag{4.3}$$

and the implicit Euler method

$$\underline{u}^{n+1} = \underline{u}^n + \Delta t \underline{f}(\underline{u}^{n+1}). \tag{4.4}$$

4.1   Order of convergence and order of consistency

There is a large number of different time integration methods. Two basic important properties necessary to describe and understand the behavior of a method are the order of convergence and the order of consistency.

Definition 1 *The local truncation error of a method is defined as the difference between the exact solution of an IVP and the numerical solution obtained after one step if exact initial data is used:*

$$l = u(t_{n+1}) - u(t_n) - \Delta t \Phi(u(t)). \tag{4.5}$$

Definition 2 *A method is called consistent of order p if for any right-hand side $f \in C^{p+1}$, the norm of the local truncation error (4.5) is $\mathcal{O}(\Delta t^{p+1})$.*

If a method is consistent, the local truncation error will converge to zero for Δt to zero or otherwise put then the method actually solves the correct problem. This property refers one time step, and one advantage is that it can be measured relatively cheap, as opposed to the global error. A more important property is the possibility whether the error after a large number of steps still has something to do with the solution.

Definition 3 *A method is called convergent of order p if for any right-hand side $f \in C^{p+1}$, the norm of the error $e_n := u(t_n) - u_n$ is $\mathcal{O}(\Delta t^p)$.*

The order of convergence is an important property of any method and the idea is that higher-order methods will allow for higher time steps for a given error tolerance, but a smaller cost per unit step.

It should be noted that due to the use of Landau symbols in the definition, consistency and convergence are defined only for $\Delta t \to 0$, whereas numerical calculations are carried out for Δt away from zero. Therefore, consistency alone is not sufficient to obtain convergence, and we need a notion of stability in addition to that.

4.2   Stability

Roughly speaking, a method is called stable, if it is robust with respect to the initial data. This implies that rounding errors do not accumulate or more precise that the error remains bounded for t to infinity for a fixed time step size. This is very difficult to establish for a general right-hand side, which is why a large number of stability properties exist for special equations.

4.2.1 The linear test equation, A- and L-stability

Important insights can be gained already in the scalar case using the Dahlquist test equation

$$\frac{d}{dt}u = \lambda u, \; u(0) = 1. \tag{4.6}$$

The exact solution is $u(t) = e^{\lambda t} = e^{Re\lambda t}e^{Im\lambda t}$. For a λ with negative real part, the exact solution decays to zero for t to infinity. Consequently, a method is stable if the numerical solution to this problem remains bounded. This depends on the step size Δt. If we consider only schemes with fixed step sizes, the set of all complex numbers $\Delta t\lambda$ for which the method is stable is called the stability region of the method. This stability region differs widely from method to method; see [122]. For the explicit Euler method, it is easy to show that the stability region is the circle around -1 with radius 1, whereas for the implicit Euler method it is the complete complex plane minus the circle around 1 with radius 1.

Since for a λ with positive real part, the solution is unbounded, it cannot be expected that the error remains bounded. Therefore, it is the left half of the complex plane that is important in the evaluation of the stability of a numerical method. This gives rise to the notion of A-stability given below.

Definition 4 *A scheme is called A-stable, if the stability region contains the left complex half plane.*

From the stability regions described above, it follows that the implicit Euler method is A-stable, whereas the explicit Euler method is not. Generally speaking, explicit methods are not A-stable, whereas implicit methods can be.

This property is rather strong, but turns out to be insufficient for the solution of CFD problems, because A-stability is not able to prevent oscillations due to very high amplitude eigenmodes. Essentially, it ignores the impact of $e^{Im\lambda t}$ on the solution. Therefore, the following more strict property is required. Here, $R(z)$ is the stability function of a onestep method for the test equation, defined by

$$u^{n+1} = R(\Delta t\lambda)u^n. \tag{4.7}$$

In other words, the stability region of a onestep method is the set of all z with $|R(z)| \leq 1$.

Definition 5 *A scheme is called L-stable, if it is A-stable and furthermore, the stability function satisfies*

$$\lim_{z\to\infty} R(z) = 0.$$

These stability properties are derived using a scalar linear test equation and the question is what the relevance is for more complex equations. For a linear system with constant coefficients

$$\frac{d}{dt}\underline{u}(t) = \mathbf{A}\underline{u},$$

it can be proved that if an RK method is A-stable, then it is unconditionally contractive in the 2-norm, if the matrix $\mathbf{A} \in \mathbb{R}^{m \times m}$ is normal and the real part of its eigenvalues is negative. For a nonnormal matrix $\mathbf{A}$, this is not true and we must expect a more severe constraint on the time step.

For nonlinear equations, even less can be said. Therefore, more general stability properties aimed at nonlinear equations like AN-stability have been suggested [52]. However, A- and L-stability seem to be sufficient for the cases discussed here.

4.2.2   TVD stability and SSP methods

The stability properties discussed so far are aimed at general equations and do not take into account special properties of the IVPs considered here, which arise from the space discretization of flow problems. As remarked in the last chapter (see section 3.6.1), there is an interesting form of stability that can be used to prove convergence of finite volume methods. This is the so-called strong stability [112], originally suggested in [247] and [248] under the name of TVD-stability. Since then, a large number of articles have been published on this topic; see for example the recent review articles [111] and [156]. To define strong stability, we assume that there exists a value Δt_{EE}, such that the explicit Euler method is stable for a certain class of functions $\underline{f}$ in the following sense for all $0 \leq \Delta t \leq \Delta t_{EE}$:

$$\|\underline{u}^n + \Delta t \underline{f}(\underline{u}^n)\| \leq \|\underline{u}^n\| \tag{4.8}$$

for an appropriate norm $\| \cdot \|$. This is in line with the requirement that the total variation does not increase with time, which corresponds to our definition of TVD schemes. The following definition carries this over to higher-order time integration methods.

Definition 6 *An s-step method is called strong stability preserving (SSP) on a certain class of functions with strong stability constant $c > 0$, if it holds that*

$$\|\underline{u}^{n+1}\| \leq \max\{\|\mathbf{u}^n\|, \|\mathbf{u}^{n-1}\|, ..., \|\mathbf{u}^{n+1-s}\|\}$$

for any time step $\Delta t \leq c\Delta t_{EE}$.

This can be related to A-stability in that a method is A-stable, if it is unconditionally SSP for the linear test equation with the real part of $\lambda \in \mathbb{C}^-$ in the norm $| \cdot |$. The crucial question is now, if we can construct methods with large c. There is good news that the implicit Euler method is unconditionally

SSP. However, for any general linear method of order greater than 1, the SSP constant is finite. Furthermore, explicit RK methods or SDIRK methods that are SSP can have an order of at most 4. A more precise bound has not been proved so far, however, convincing numerical evidence obtained by an optimization software shows the following results [156]. For a second-order SDIRK method with s stages, the bound is $2s$, for third-order SDIRK, it is $s - 1 + \sqrt{s^2 - 1}$. For higher orders, the bounds obtained are roughly speaking $2s$. For linear multistep methods of order greater than 1, the SSP constant is smaller or equal to 2.

These SSP coefficients are well below what is needed to make implicit methods competitive. Therefore, when using an implicit method, we do not know for sure that our numerical method is TVD. This does not seem to matter at all in practice and suggests that the SSP property is too strong to require. Nevertheless, for problems where explicit methods are faster than implicit ones, SSP methods should be preferred to be on the safe side.

4.2.3 The CFL condition, von Neumann stability analysis and related topics

The stability conditions so far have been derived using ordinary differential equations only. If we additionally take into account that we consider discretized partial differential equations, more insight can be gained. Furthermore, the maximal stable time step Δt_{EE} for the explicit Euler method used in the SSP theory can be determined. For a more detailed overview than here, the books [138, 301, 119] are very useful.

First of all, for the hyperbolic inviscid terms, we know that for a scheme to be stable, it has to satisfy the Courant-Friedrichs-Levy (CFL) condition. This states that the domain of dependence of the solution has to contain the domain of dependence of the numerical method [65]. This is illustrated in Figure 4.1. The CFL condition is thus a nonlinear sufficient condition for stability on the time step.

It is automatically satisfied by implicit schemes, regardless of the time step. For finite volume methods for one-dimensional equations with explicit time integration, we obtain the constraint

$$\Delta t < CFL_{\max} \cdot \frac{\Delta x}{\max\limits_{k,|\mathbf{n}|=1} |\lambda_k(\mathbf{u}, \mathbf{n})|}, \tag{4.9}$$

with the maximal CFL number $CFL_{\max}$ depending on the method; for example 1.0 for the explicit Euler method. The λ_k are the eigenvalues (2.27) of the Jacobian of the inviscid flux.

In practice, this is often used to compute the time step, even for implicit methods. To this end, one defines a CFL number $CFL \leq CFL_{\max}$ and

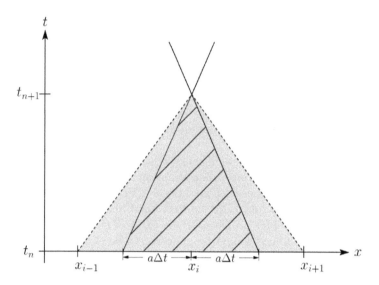

FIGURE 4.1: Illustration of the CFL condition for a linear equation: the exact solution in the point (x_i, t_{n+1}) can be influenced by the points in the shaded area. The numerical domain of dependence (the grey area) has to contain the shaded area for the scheme to be stable, resulting in a constraint on $\Delta t/\Delta x$. (Credit: V. Straub.)

computes

$$\Delta t = CFL \cdot \frac{\Delta x}{\max_{k,|\mathbf{n}|=1} |\lambda_k(\mathbf{u}, \mathbf{n})|}.$$

To obtain necessary conditions, von Neumann stability analysis is used. This is, besides the energy method, the standard tool for the stability analysis of discretized partial differential equations. A thorough discussion of the technique can be found in [119]. We give a brief explanation in section 5.6.7.6 and mainly summarize results here. The technique considers a linear equation with periodic boundary conditions. Then, a linear discretization on an equidistant mesh is applied, and Fourier data is inserted into the approximation, which decouples the system. The discretization is stable in L_2 if no Fourier component is amplified, typically leading to conditions on Δt and Δx. For an upwind discretization combined with the explicit Euler method, this results in the condition (4.9). This means that the CFL condition is sharp for linear problems where it is both necessary and sufficient. Unfortunately, this technique becomes extremely complicated when looking at unstructured grids, making it less useful for these.

For nonlinear equations, the analysis is applied to a linearized version. We then obtain results that are not identical to the CFL condition. To obtain a sharp condition, additional stability constraints are needed. For the case of the viscous Burger's equation, this has been demonstrated in [185]. For general equations, no complete stability analysis is known.

The assumption of periodic boundary conditions is most often not satisfied. In practice, one applies the results discussed here, which works rather well, but it has to be pointed out that robustness is an issue for turbulent flow computations. The analysis of fully discrete schemes without periodic boundary conditions is a mostly open research question, at least in the nonlinear case.

If additional parabolic viscous terms are present, like in the Navier-Stokes equations, the stability constraint changes. For a pure diffusion equation, resp. the linear heat equation discretized with central differences and explicit Euler in time, one can use a von Neumann stability analysis and the CFL condition about the domains of dependence to obtain the necessary and sufficient condition

$$\Delta t < \frac{\Delta x^2}{2\nu}, \tag{4.10}$$

where ν is the diffusion coefficient. Here, the dependence on the mesh width is quadratic and thus, this condition is much more severe for fine grids than the condition (4.9) for hyperbolic equations.

One way of obtaining a stable time integration method for the Navier-Stokes equations is to require the scheme to satisfy both (4.9) and (4.10). However, this is too severe, as seen in practice and by a more detailed analysis. For example, the restriction on the time step in theorem 4 for the equation $u_t + \nabla \cdot f(u) = \nu \Delta u$ is of the form

$$\Delta t < \frac{\alpha^3 \Delta x^2}{\alpha \max\limits_{k, |\mathbf{n}|=1} |\lambda_k(\mathbf{u}, \mathbf{n})| \Delta x + \nu}. \tag{4.11}$$

Here, α is a grid-dependent factor that is roughly speaking closer to one the more regular the grid is (see [214] for details). A similar bound has been found in praxis to be useful for the Navier-Stokes equations [44]:

$$\Delta t < \frac{\Delta x^2}{\max\limits_{k, |\mathbf{n}|=1} |\lambda_k(\mathbf{u}, \mathbf{n})| \Delta x + 2\epsilon}.$$

In the case of a DG method, both time step constraints additionally depend on the order of the polynomial basis in that there is an additional factor $1/(2N+1)$ in condition (4.9) and of $1/N^2$ in (4.10); see [62]. For higher orders, this is not a sharp constraint, but a guideline that works in practice. The dependence of the stability constraint on the choice of the viscous flux has been considered in [151].

In more than one space dimensions, the situation is, as expected, more difficult, in particular for arbitrary polygonal cells. It is not obvious which value to choose in a cell for Δx. A relation to determine a locally stable time step in a cell Ω that works on both structured and unstructured grids for finite volume schemes is the following:

$$\Delta t = \sigma \frac{|\Omega|}{\lambda_{c,1} + \lambda_{c,2} + \lambda_{c,3} + 4(\lambda_{v,1} + \lambda_{v,2} + \lambda_{v,3})}.$$

Here, $\lambda_{c,i} = (|v_i| + c)|s_i|$ and

$$\lambda_{v,i} = \max\left(\frac{4}{3\rho}, \frac{\gamma}{p}\right) \frac{\mu}{Pr} \frac{|s_i|^2}{|\Omega|},$$

where s_i is a projection of the control volume on the planes orthogonal to the x_i direction [293]. The parameter σ corresponds to a CFL number and has to be determined in practice for each scheme, due to the absence of theory. Finally, a globally stable time step is obtained by taking the minimum of the local time steps.

4.3   Stiff problems

Obviously, implicit methods are more costly than explicit ones. The reason they are considered nevertheless are stiff problems, which can somewhat tautological be defined as problems where implicit methods are more efficient than explicit ones. Broadly speaking, this can happen since implicit schemes have better stability properties than explicit ones. If the problem is such that a stability constraint on the time step size leads to time steps much smaller than useful to resolve the physics of the system considered, the additional cost per time step is justified, making implicit schemes the method of choice. To illustrate this phenomenon, consider the system of two equations

$$x_t = -x,$$
$$y_t = -1000y$$
$$x(0) = y(0) = 1.$$

The fast scale quantity $y(t) = e^{-1000t}$ will decay extremely fast to near zero and thus, large time steps that resolve the evolution of the slow scale quantity $x(t) = e^{-t}$ should be possible; see Figure 4.2. However, a typical explicit scheme will be hampered by stability constraints where it needs to choose the time step according to the fastest scale, even though this has no influence on the solution after an initial transient phase. This example illustrates one possibility of characterizing stiffness: we have a large Lipschitz constant of our function, but additionally eigenvalues of small magnitude. In this case, the long-term behavior of the solution is much better characterized by the largest real part of the eigenvalues, instead of the Lipschitz constant itself [72]. This is connected to the logarithmic norm [252]. Another possibility of obtaining stiffness are stiff source terms, meaning source terms, that introduce physics on a different time scale than the rest of the flow. Note that whether stiffness manifests itself also depends on the initial values chosen, as well as on the maximal time considered, which is why it is better to talk of stiff problems instead of stiff equations. For example, the time scale considered could be so small that the larger time steps of an implicit method cannot pay off.

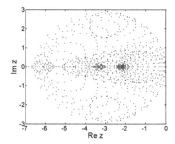 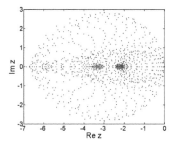

FIGURE 4.2: Solution of the stiff example equation with initial condition (10, 10) (left); Eigenvalues for a two-dimensional Navier-Stokes DG discretization with $Re = 100$, 6×6 mesh, fourth order in space (right) (credit: V. Straub).

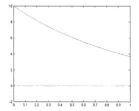 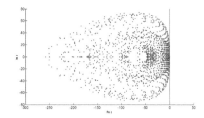

FIGURE 4.3: Eigenvalues for a two-dimensional finite volume discretization for Euler (left), and Navier-Stokes with $Re = 1000$ (right).

Flow problems have similar properties as the above system with both slow and fast scales present. For the Euler equations, there are fast-moving acoustic terms, which correspond to the largest eigenvalue. Then, there are the significantly slower convective terms, which correspond to the convective eigenvalue $|v|$. The time step size of explicit schemes will be dictated by the acoustic eigenvalue, even if the physical processes of interest usually live on the convective scale. This property becomes extreme for low Mach numbers, when the convective eigenvalue approaches zero and the quotient between these two eigenvalues becomes very large, e.g., stiffness increases.

In the case of the Navier-Stokes equations with a bounding wall, we have additionally the boundary layer. This has to be resolved with extremely fine grid spacing in normal direction, leading to cells which are several orders of magnitude smaller than for the Euler equations. For an explicit scheme, the CFL condition therefore becomes several orders of magnitude more restrictive, independent of the flow physics. Furthermore, large aspect ratios are another source of stiffness, sometimes called geometric stiffness. Therefore, implicit methods play an important role in the numerical computation of unsteady viscous flows. If no boundary layer is present, this problem is less severe and explicit methods still may be the methods of choice.

Finally, DG discretizations lead to a spectrum that hugs the imaginary axis, compare Figure 4.2 (right), making A-stability absolutely necessary. A finite volume discretization does not have this issue (see Figure 4.3). Now, different classes of time integration schemes will be presented in detail.

4.4   Backward differentiation formulas

A class of time integration schemes that belongs to the multistep methods are backward differentiation formulas (BDF). These approximate the solution by a pth order polynomial that interpolates the values $\underline{\mathbf{u}}^{n+1-k}$ at the points t_{n+1-k}, $k = 0, ..., p$. Since both $\underline{\mathbf{u}}^{n+1}$ and $\underline{\mathbf{f}}(\underline{\mathbf{u}}^{n+1})$ are unknown, it is additionally required that the polynomial fulfills the differential equation at t_{n+1}. This condition results in one nonlinear equation system to determine the new approximation $\underline{\mathbf{u}}_{n+1}$:

$$\sum_{i=1}^{p+1} \alpha_i \underline{\mathbf{u}}^{n+1-i} = \Delta t \underline{\mathbf{f}}(\underline{\mathbf{u}}^{n+1}). \tag{4.12}$$

Here, the coefficients α_i are defined by the interpolation polynomial and thus depend on the step size history.

The resulting methods are of order p, but unstable for $p > 6$. For $p = 1$, the implicit Euler method is obtained. The method BDF-2, obtained for $p = 2$, for a fixed time step size Δt is

$$\left(\frac{3}{2}\underline{\mathbf{u}}^{n+1} - \frac{4}{2}\underline{\mathbf{u}}^n + \frac{1}{2}\underline{\mathbf{u}}^{n-1} \right) = \Delta t \underline{\mathbf{f}}(\underline{\mathbf{u}}^{n+1}). \tag{4.13}$$

This method is A-stable and also L-stable. For orders greater than 2, this is impossible due to the second Dahlquist barrier. There, the methods are A(α)-stable, with α decreasing with order. As shown in Figure 4.2 (right), this is problematic. As for the SSP property, linear multistep methods of order greater than 1 have an SSP constant of 2 or smaller [111].

In the case of varying time step sizes, the coefficients of the method vary as well. In practice, this is neglected and thus, an additional time integration error is introduced. BDF methods with varying time step sizes are implemented in the well known ODE solver DASSL of Petzold et al. [48], as well as in SUNDIALs [134]. This class of schemes, in particular BDF-2, is often used in the engineering community, with the reason given that you get a second-order A-stable method that needs only one nonlinear system.

4.5   Runge-Kutta methods

By far, the most widely used class of onestep methods are the Runge-Kutta (RK) methods. An s-stage RK method is derived by integrating the IVP (4.1) over $[t_n, t_{n+1}]$ to

$$\underline{u}(t_{n+1}) = \underline{u}(t_n) + \int_{t_n}^{t_{n+1}} \underline{f}(t, \underline{u}(t))dt.$$

This integral is now approximated using a quadrature rule

$$I = \int_0^1 \underline{f}(\tau)d\tau \approx \sum_{i=1}^s b_i \underline{f}(c_i)$$

with weights b_i and nodes $c_i \in [0, 1], i = 1, ..., s$. We thus obtain

$$\underline{u}(t_{n+1}) \approx \underline{u}_{n+1} = \underline{u}_n + \Delta t_n \sum_{i=1}^s b_i \underline{f}(t_n + c_i \Delta t_n, \underline{u}(t_n + c_i \Delta t_n)).$$

This hasn't really helped, since the values $\underline{u}(t_n + c_i \Delta t_n)$ are unknown as well. The crucial idea is now to use another quadrature formula with possibly different weights, but the same nodes to obtain these values:

$$\underline{u}(t_n + c_i \Delta t_n) \approx \mathbf{U}_i = \underline{u}_n + \Delta t_n \sum_{j=1}^s a_{ij} \underline{f}(t_n + c_j \Delta t_n, \mathbf{U}_j).$$

The $\mathbf{U}_i$ are called stage values and correspond to approximations of the solution at $t_n + c_i \Delta t_n$, whereas

$$\mathbf{k}_i = \underline{f}(t_n + c_i \Delta t_n, \mathbf{U}_i) \tag{4.14}$$

are called stages derivatives. Summarizing, a general s-stage RK method written in terms of the stage derivatives is given by

$$\mathbf{k}_i = \underline{\mathbf{f}}(t_n + c_i \Delta t_n, \underline{\mathbf{u}}^n + \Delta t \sum_{j=1}^{i} a_{ij} \mathbf{k}_j), \quad i = 1, ..., s \qquad (4.15)$$

$$\underline{\mathbf{u}}^{n+1} = \underline{\mathbf{u}}^n + \Delta t \sum_{i=1}^{s} b_i \mathbf{k}_i.$$

For the autonomous system (4.2), this simplifies to

$$\mathbf{k}_i = \underline{\mathbf{f}}(\underline{\mathbf{u}}^n + \Delta t_n \sum_{j=1}^{i} a_{ij} \mathbf{k}_j), \quad i = 1, ..., s, \qquad (4.16)$$

$$\underline{\mathbf{u}}^{n+1} = \underline{\mathbf{u}}^n + \Delta t_n \sum_{i=1}^{s} b_i \mathbf{k}_i.$$

The coefficients of the scheme can be written in a compact form using the vectors $\mathbf{c} = (c_1, ..., c_s)^T$, $\mathbf{b} = (b_1, ..., b_s)^T$, and the matrix $\mathbf{A} = (a_{ij})_{ij} \in \mathbb{R}^{s \times s}$. These can be combined in the so-called Butcher array, which is the standard way of writing down the coefficients of an RK scheme:

$$
\begin{array}{c|ccc}
c_1 & a_{11} & \cdots & a_{1s} \\
\vdots & \vdots & \cdots & \vdots \\
c_s & a_{s1} & \cdots & a_{ss} \\
\hline
& b_1 & \cdots & b_s
\end{array}
=
\begin{array}{c|c}
\mathbf{c} & \mathbf{A} \\
\hline
& \mathbf{b}^T
\end{array}
$$

An explicit RK method is obtained, if the matrix $\mathbf{A}$ is strictly lower triangular, otherwise the method is implicit. The stability function is the rational function

$$R(z) = \frac{\det(\mathbf{I} - z\mathbf{A} + z\mathbf{e}\mathbf{b}^T)}{\det(\mathbf{I} - z\mathbf{A})}, \qquad (4.17)$$

where $\mathbf{e} \in \mathbb{R}^s$ is the vector of all ones. This can be seen by looking at the equation system obtained when solving the linear test equation and solving that using Cramer's rule. For an A-stable method, if either the first column in $\mathbf{A}$ coincides with $\mathbf{c}$ or the last row in $\mathbf{A}$ with $\mathbf{b}^T$, L-stability is obtained. This class of schemes is called *stiffly accurate* and is popular for the solution of differential algebraic equations (DAEs).

The maximal order of an s-stage RK method is $2s$, which is obtained by the so-called Gauß-Legendre methods. One way to obtain RK methods of a certain order is to use Taylor expansions and coefficient matching. To keep track of the high order mixed derivatives quickly becomes overwhelming. Therefore, the concept of Butcher trees is used [121]. Nevertheless, this results in a huge number of equations to determine the coefficients. Furthermore, because products of coefficients appear, they are nonlinear. Therefore, Butcher devised simplifying conditions that allow to solve these nonlinear systems even for high orders. In this way, it is possible to find RK methods with different orders, degrees of freedom, and stability properties.

Several time integration schemes are widely used, for example the Crank-Nicholson scheme, also called trapezoidal rule, which is A-stable:

$$\mathbf{k}_1 = \underline{\mathbf{f}}(t_n, \underline{\mathbf{u}}^n),$$

$$\mathbf{k}_2 = \underline{\mathbf{f}}\left(t_n + \Delta t_n, \underline{\mathbf{u}}^n + \Delta t_n\left(\frac{1}{2}\mathbf{k}_1 + \frac{1}{2}\mathbf{k}_2\right)\right),$$

$$\underline{\mathbf{u}}^{n+1} = \underline{\mathbf{u}}^n + \Delta t\left(\frac{1}{2}\mathbf{k}_1 + \frac{1}{2}\mathbf{k}_2\right).$$

4.5.1 Explicit Runge-Kutta methods

0	0	0	0	0
c_2	a_{21}	0	0	0
$\vdots$	$\vdots$	$\ddots$	$\ddots$	0
c_s	a_{s1}	$\cdots$	a_{ss-1}	0
	b_1	$\cdots$	$\cdots$	b_s

TABLE 4.1: Butcher array of an explicit RK method.

For explicit Runge-Kutta methods, the Butcher array (see Table 4.1) has a strictly lower triangular form. Thus, the denominator in the stability function (4.17) is 1 and it becomes a polynomial of degree s for an s-stage method. Therefore, the stability region is bounded and thus, these methods cannot be A-stable. There is a variety of popular explicit RK methods, in particular the explicit Euler method (4.3), the improved Euler method, which is a second-order explicit RK scheme or the classical RK method of fourth order.

Regarding order, since a large number of the a_{ij} is zero, there are significantly less degrees of freedom in the design of the method. Nevertheless, the order of an s-stage method can be up to s for $s \leq 4$. Beyond that, the maximal order of an s-stage explicit RK method is smaller than s.

As for the choice of method, explicit RK methods are useful if the stability constraint is not severe with regards to the time scale we are interested in. Now, if oscillations are a problem, SSP methods should be employed. This is essentially never the case for finite volume methods, where the inherent diffusion is quite large. However, for high order methods like DG, oscillations can become a problem. The coefficients of k-stage, kth-order methods with an optimal SSP constant of 1 can be found in the appendix table B.1. These methods are called SSP2 and SSP3. Note that SSP2 is the improved Euler method, also called Heun's method.

If oscillations are not an issue, low storage explicit RK (LSERK) methods are a good option that need only two vectors for the solution [54]. These are given by

$$\underline{u} = \underline{u}^n$$
$$\underline{k} = \underline{0}$$
$$i \in [1, ..., s] : \begin{cases} \underline{k} = a_i \, \underline{k} + \Delta t \underline{f}(\underline{u}) \\ \underline{u} = \underline{u} + b_i \, \underline{k} \end{cases} \tag{4.18}$$
$$\underline{u}^{n+1} = \underline{u}$$

Suitable coefficients can be found in the appendix table B.3.

4.5.2   Fully implicit RK methods

In contrast to multistep schemes, no bound on the order for A-stable schemes has been proved or observed for implicit Runge-Kutta (IRK) methods. This is because the stability function is rational and therefore, the stability region can be unbounded. In a fully implicit Runge-Kutta method, the Butcher array is filled and thus, a nonlinear equation system with $s \cdot m$ unknowns has to be solved at every step. There are three main classes of interest, Gauß-Legendre, Radau IIA, and Gauß-Lobatto IIIC. All of them are A-stable. The first class has the highest order, $2s$, being based on Gauß-quadrature. However, these methods are not L-stable, which reduces their usefulness for CFD problems. The methods in the other two classes, however, are stiffly accurate, and therefore L-stable. These methods also have the SBP-SAT property in time [230]. The coefficients can be found in the appendix table B.4 for the Radau methods and in appendix tables B.5 and B.6 for the Gauß-Lobatto methods. The use of Radau IIA methods was compared to Gauß-Legendre methods in [140]. The former were found to have a better performance, also compared to BDF methods and the later discussed SDIRK methods.

For Gauß-Lobatto methods, the quadrature nodes include the endpoints. The class of Gauß-Lobatto IIIc methods arises in fact also, when one uses DG-SEM for integration in time.

4.5.3   DIRK methods

Several methods have been suggested to avoid the huge nonlinear systems in IRK methods, for example SIRK methods where $\mathbf{A}$ is diagonalizable. We restrict ourselves to so-called diagonally implicit Runge-Kutta (DIRK) methods. Given coefficients a_{ij} and b_i, such a method with s stages can be written as

$$\mathbf{k}_i = \underline{f}(t_n + c_i \Delta t, \underline{u}^n + \Delta t \sum_{j=1}^{i} a_{ij} \mathbf{k}_j), \quad i = 1, ..., s \tag{4.19}$$

$$\underline{u}^{n+1} = \underline{u}^n + \Delta t \sum_{i=1}^{s} b_i \mathbf{k}_i.$$

Thus, all entries of the Butcher array in the strictly upper triangular part are zero. If additionally all values on the diagonal of $\mathbf{A}$ are identical, the scheme is called singly diagonally implicit (SDIRK). The stability function of a DIRK method is given by

$$R(z) = \frac{\det(\mathbf{I} - z\mathbf{A} + z\mathbf{e}\mathbf{b}^T)}{\Pi_{i=1}^s(1 - za_{ii})}.$$

Regarding order, an s-stage SDIRK method can be of order $s+1$. This is a bit surprising, since there is only one additional degree of freedom compared to an explicit method. To obtain L-stability, one chooses the last row in $\mathbf{A}$ to coincide with $\mathbf{b}$. This reduces the order to at most s. Finally, there is the class of ESDIRK schemes, where the first step of the RK schemes is explicit. The butcher arrays of SDIRK and ESDIRK methods are illustrated in Table 4.2.

α	α	0	0	0
c_2	a_{21}	α	0	0
$\vdots$	$\vdots$	$\ddots$	$\ddots$	0
c_s	a_{s1}	$\cdots$	a_{ss-1}	α
	b_1	$\cdots$	$\cdots$	b_s

0	0	0	0	0
c_2	a_{21}	α	0	0
$\vdots$	$\vdots$	$\ddots$	$\ddots$	0
c_s	a_{s1}	$\cdots$	a_{ss-1}	α
	b_1	$\cdots$	$\cdots$	b_s

TABLE 4.2: Butcher array of an SDIRK method (left) and an ESDIRK method (right)

The point about DIRK schemes is that the computation of the stage vectors is decoupled and instead of solving one nonlinear system with sm unknowns, the s nonlinear systems (4.19) with m unknowns have to be solved. This corresponds to the sequential application of several implicit Euler steps in two possible ways. With the starting vectors

$$\mathbf{s}_i = \underline{\mathbf{u}}^n + \Delta t \sum_{j=1}^{i-1} a_{ij}\mathbf{k}_j, \tag{4.20}$$

we can solve for the stage derivatives

$$\mathbf{k}_i = \underline{\mathbf{f}}(\mathbf{s}_i + \Delta t a_{ii}\mathbf{k}_i) \tag{4.21}$$

or we define the stage values via

$$\mathbf{U}_i = \underline{\mathbf{u}}^n + \Delta t \sum_{j=1}^{i} a_{ij}\mathbf{k}_j, \tag{4.22}$$

which implies

$$\mathbf{k}_i = \underline{\mathbf{f}}(\mathbf{U}_i)$$

and then solve the equation

$$\mathbf{U}_i = \mathbf{s}_i + \Delta t a_{ii}\underline{\mathbf{f}}(\mathbf{U}_i). \tag{4.23}$$

In the autonomous case, equations (4.21) and (4.23) correspond to one step of the implicit Euler method with starting vector $\mathbf{s}_i$ and time step $a_{ii}\Delta t$. If (4.23) is employed, the stage value $\mathbf{k}_i$ is then obtained by

$$\mathbf{k}_i = (\mathbf{U}_i - \mathbf{s}_i)/(a_{ii}\Delta t),$$

which avoids a costly and for stiff problems error prone evaluation of the right-hand side [246]. The major difference between the two formulations is that the iterative solver used to solve one of the equations will produce errors in either the stage values or the stage derivatives, leading to a different error propagation [215]. Finally, we apply the formula

$$\underline{\mathbf{u}}^{n+1} = \underline{\mathbf{u}}^n + \Delta t \sum_{i=1}^{s} b_i \mathbf{k}_i.$$

Note that this is not necessary for stiffly accurate methods that work with stage values, since there $\underline{\mathbf{u}}^{n+1} = \mathbf{U}_s$. Therefore, we will only consider stiffly accurate methods, since they have beneficial stability properties and less computational costs.

All in all, the application of a DIRK scheme corresponds to the solution of s nonlinear systems of the form (4.19) per time step and therefore, the intuitive thought is not to consider this class of schemes, since they are inefficient compared to BDF methods. However, several facts lead to a positive reevaluation of these schemes. First of all, a_{ii} is typically smaller than one and thus the equation systems are easier to solve than a system arising from an implicit Euler or BDF2 discretization with the same Δt. Second, if a time-adaptive strategy is used, the higher order leads to larger time steps compared to the implicit Euler method and therefore fewer time steps are needed to integrate from t_0 to t_{end}, leading to a smaller total number of nonlinear systems to be solved. Finally, since we have a sequence of nonlinear systems which continuously depend on each other, results from previous systems can be exploited to speed up the solution of later systems. This will be explained in the following chapter.

Additionally, it is usually possible to give a second set of coefficients $\hat{b}$, which define for the otherwise identical Butcher tableau a method of lower order. This can be used for the cheap estimation of the local time integration error, as will be explained later. Furthermore, so-called dense output formulas can be found. These are designed to deliver values of the solution inside the interval $[t_n, t_{n+1}]$, once $\underline{\mathbf{u}}^{n+1}$ has been computed. However, they can be used to extrapolate the solution into the future to obtain starting values for the iterative solution methods for the nonlinear equation systems. This will also be discussed in the next chapter. The formulas itself are given by

$$\underline{\mathbf{u}}(t_n + \theta \Delta t) = \underline{\mathbf{u}}^n + \Delta t \sum_{i=1}^{s} b_i^*(\theta)\underline{\mathbf{f}}(\underline{\mathbf{u}}^{(i)}), \qquad (4.24)$$

where the factors $b_i^*(\Theta)$ are given by

$$b_i^*(\theta) = \sum_{j=1}^{p^*} b_{ij}^* \theta^j, \quad b_i^*(\theta = 1) = b_i, \tag{4.25}$$

with b_{ij}^* being a set of method-dependent coefficients and p^* the order of the formula.

One example for an SDIRK scheme is the method of Ellsiepen [84], which is of second order with an embedded method of first order (see Table B.7). A method of third order with an embedding of second order was developed by Cash (see Table B.8). In the context of solid mechanics, it has been demonstrated by Hartmann that these methods are competitive [125].

Regarding ESDIRK schemes, this allows to have a stage order of 2, but also means that the methods cannot be algebraically stable. The use of these schemes in the context of compressible Navier-Stokes equations was analyzed by Bijl et al. in [23] where they were demonstrated to be more efficient than BDF methods for engineering accuracies. They suggested the six-stage method ESDIRK4 of fourth order with an embedded method of third order and the four-stage method ESDIRK3 of third order with an embedded method of second order, both designed in [155]. The coefficients can be obtained from the diagrams (B.9) and (B.10), and (B.11) and (B.12), respectively for the dense output formulas. This result about the comparative efficiency of BDF-2 and ESDIRK4 was later confirmed by Jothiprasad et al. for finite volume schemes [149] and Wang and Mavriplis for a DG discretization of unsteady Euler flow [298].

The use of a first explicit stage $\mathbf{k}_1 = \mathbf{f}(\underline{u}^n)$ in a stiffly accurate method allows to reuse the final stage derivative $\mathbf{k}_s$ from the last time step in the first stage, since this is equal to $\mathbf{f}(\underline{u}^{n-1} + \Delta t_{n-1} \sum_{i=1}^s b_i \mathbf{k}_i)$. Besides saving a modest amount of computing time, this is in particular advisable for stiff problems with a large Lipschitz constant. There, small errors in the computation of stage values may lead to large errors in evaluated functions, whereas small errors in the computed stage derivative are just that. The drawback here is that the methods then no longer are one-step methods in the strict sense, since we have to store the last stage derivative.

Finally, a word of caution: DIRK methods, being a sequence of implicit Euler steps have only order 1 at the individual stages. This can cause order reduction for extremely stiff problems [181]. In the examples considered here, we have not observed this, though.

4.5.4 Additive Runge-Kutta methods

An idea that came up in the 1980s is to use different RK methods for different terms in the equation [64], which was introduced to the CFD community independently in the famous Jameson-Schmidt-Turkel paper [146]. There, explicit additive methods are suggested to be used as smoothers in his multigrid methods and to treat the convective and diffusive terms differently. This will be explained later. Another point about this are implicit-explicit methods (IMEX) as designed in [155], which treat nonstiff terms explicitly and stiff terms implicitly. We will here represent the idea for two terms, but the extension to N terms is straightforward. Consider the ODE

$$\frac{d}{dt}\underline{u}(t) = \underline{f}(t, \underline{u}(t)), \quad \underline{u}(t_0) = \underline{u}_0, \quad t \in [t_0, t_{\text{end}}], \tag{4.26}$$

with

$$\underline{f}(t, \underline{u}(t)) = \underline{f}_1(t, \underline{u}(t)) + \underline{f}_2(t, \underline{u}(t)).$$

The method is then applied as

$$\mathbf{U}_i = \underline{u}^n + \Delta t \sum_{\nu=1}^{2} \sum_{j=1}^{i} a_{ij}^{(\nu)} \underline{f}_\nu(\mathbf{U}_j), \quad i = 1, ..., s \tag{4.27}$$

$$\underline{u}^{n+1} = \underline{u}^n + \Delta t \sum_{\nu=1}^{2} \sum_{i=1}^{s} b_i^{(\nu)} \underline{f}_\nu(\mathbf{U}_i).$$

4.6 Rosenbrock-type methods

To circumvent the solution of nonlinear equation systems, so-called Rosenbrock methods can be used, which are also referred to as Rosenbrock-Wanner (ROW) methods (see [122] or the German volume [264]). The idea is to linearize an s-stage DIRK scheme, thus sacrificing some stability properties, as well as accuracy, but reducing the computational effort in that per time step, s linear equation systems with the same system matrix and different right-hand sides have to be solved. Therefore, this class of schemes is also sometimes referred to as linearly implicit or semi-implicit. Note that semi-implicit is used in an ambiguous way in different contexts and therefore shouldn't be used. In [148], Rosenbrock methods are compared to SDIRK methods in the context of the incompressible Navier-Stokes equations and found to be competitive, if not superior. The use of Rosenbrock methods in the context of compressible flow problems is so far quite rare. One example is the work of St.-Cyr et al. in the context of DG [260], as well as the study [45] for FV methods.

For the derivation of the schemes, we start by linearizing formula (4.19) around $\mathbf{s}_i$ (4.20) to obtain

$$\mathbf{k}_i \approx \underline{f}(\mathbf{s}_i) + \Delta t a_{ii} \frac{\partial \underline{f}(\mathbf{s}_i)}{\partial \underline{u}} \mathbf{k}_i.$$

To avoid a recomputation of the Jacobian $\frac{\partial \underline{f}(\mathbf{s}_i)}{\partial \underline{u}}$ at every stage, it is replaced by $\mathbf{J} = \frac{\partial \underline{f}(\underline{u}^n)}{\partial \underline{u}}$. Finally, to gain more freedom in the definition of the method, linear combinations of $\Delta t \mathbf{J} \mathbf{k}_i$ are added to the last term. Since the linearization procedure can be interpreted as performing just one Newton step at every stage of the DIRK method instead of a Newton loop, the added terms correspond roughly to choosing that as the initial guess for the first Newton iteration. If instead of the exact Jacobian, an approximation $\mathbf{W} \approx \mathbf{J}$ is used, we obtain so-called W-methods. Since for the systems considered here, exact solutions are impossible, we will consider W-methods only. If the linear system is solved using a Krylov subspace method (see section 5.8), the scheme is called a Krylov-ROW method. This is for example implemented in the code ROWMAP [300].

We thus obtain an s-stage ROW method with coefficients a_{ij}, γ_{ij}, and b_i in the form

$$
\begin{aligned}
(\mathbf{I} - \gamma_{ii}\Delta t \mathbf{W})\mathbf{k}_i &= \underline{\mathbf{f}}(\mathbf{s}_i) + \Delta t \mathbf{W} \sum_{j=1}^{i-1} \gamma_{ij}\mathbf{k}_j, \quad i = 1, ..., s \\
\mathbf{s}_i &= \underline{\mathbf{u}}^n + \Delta t \sum_{j=1}^{i-1} a_{ij}\mathbf{k}_j, \quad i = 1, ..., s \qquad (4.28) \\
\underline{\mathbf{u}}^{n+1} &= \underline{\mathbf{u}}^n + \Delta t \sum_{i=1}^{s} b_i \mathbf{k}_i.
\end{aligned}
$$

Here, the coefficients a_{ij} and b_i correspond to those of the DIRK method and the γ_{ii} are the diagonal coefficients of that, whereas the offdiagonal γ_{ij} are additional coefficients.

In the case of a nonautonomous equation (4.1), we obtain an additional term $\Delta t \gamma_i \partial_t \underline{\mathbf{f}}(t_n, \underline{\mathbf{u}}_n)$ with $\gamma_i = \sum_{j=1}^{i} \gamma_{ij}$ on the right-hand side of (4.28) and thus:

$$
\begin{aligned}
(\mathbf{I} - \gamma_{ii}\Delta t \mathbf{W})\mathbf{k}_i &= \underline{\mathbf{f}}(t_n + a_i\Delta t, \mathbf{s}_i) + \Delta t \mathbf{W} \sum_{j=1}^{i-1} \gamma_{ij}\mathbf{k}_j + \Delta t \gamma_i \partial_t \underline{\mathbf{f}}(t_n, \underline{\mathbf{u}}_n), \\
&\qquad i = 1, ..., s \\
\mathbf{s}_i &= \underline{\mathbf{u}}^n + \Delta t \sum_{j=1}^{i-1} a_{ij}\mathbf{k}_j, \quad i = 1, ..., s \qquad (4.29) \\
\underline{\mathbf{u}}^{n+1} &= \underline{\mathbf{u}}^n + \Delta t \sum_{i=1}^{s} b_i \mathbf{k}_i,
\end{aligned}
$$

with $a_i = \sum_{j=1}^{i-1} a_{ij}$.

Order trees for Rosenbrock methods can be derived using the same techniques as for RK methods and the result is very similar to that for DIRK methods. For W-methods, additional order conditions are needed to avoid a loss of order.

Generally speaking, Rosenbrock methods are less accurate than SDIRK methods, which will result later in the time step selector choosing smaller time steps for Rosenbrock methods. The stability function of a Rosenbrock method is that of a DIRK method where the matrix in the butcher array has entries $a_{ij} + \gamma_{ij}$. Since A- and L-stability are linear concepts, it is not difficult to construct W-methods with that property.

The efficient implementation is done using a set of auxiliary variables

$$\mathbf{U}_i = \Delta t \sum_{j=1}^{i} \gamma_{ij} \mathbf{k}_j$$

and thus circumvents the matrix-vector multiplication in the previous formulation (4.28). Using the identity

$$\mathbf{k}_i = \frac{1}{\Delta t} \left(\frac{1}{\gamma_{ii}} \mathbf{U}_i - \sum_{j=1}^{i-1} c_{ij} \mathbf{U}_j \right)$$

with coefficients c_{ij} explained below, we obtain

$$(\mathbf{I} - \gamma_{ii}\Delta t \mathbf{W})\mathbf{U}_i = \Delta t \gamma_{ii} \underline{\mathbf{f}}(\hat{\mathbf{s}}_i) + \gamma_{ii} \sum_{j=1}^{i-1} c_{ij}\mathbf{U}_j, \quad i = 1, ..., s, \quad (4.30)$$

$$\hat{\mathbf{s}}_i = \underline{\mathbf{u}}^n + \sum_{j=1}^{i-1} \tilde{a}_{ij}\mathbf{U}_j, \quad i = 1, ..., s,$$

$$\underline{\mathbf{u}}^{n+1} = \underline{\mathbf{u}}^n + \sum_{i=1}^{s} m_i \mathbf{U}_i.$$

The relation between the two sets of coefficients is the following:

$$\mathbf{C} = \text{diag}(\gamma_{11}^{-1}, ..., \gamma_{ss}^{-1}) - \mathbf{\Gamma}^{-1}, \qquad \tilde{\mathbf{A}} = \mathbf{A}\mathbf{\Gamma}^{-1}, \qquad \mathbf{m}^T = \mathbf{b}^T\mathbf{\Gamma}^{-1},$$

where $\mathbf{\Gamma} = (\gamma_{ij})_{ij}$. Again, in the nonautonomous case, we obtain an additional term on the right-hand side and thus:

$$(\mathbf{I} - \gamma_{ii}\Delta t \mathbf{W})\mathbf{U}_i = \Delta t \gamma_{ii} \underline{\mathbf{f}}(t_n + a_i\Delta t, \hat{\mathbf{s}}_i) + \gamma_{ii} \sum_{j=1}^{i-1} c_{ij}\mathbf{U}_j$$

$$+\Delta t^2 \gamma_{ii}\gamma_i \partial_t \underline{\mathbf{f}}(t_n, \underline{\mathbf{u}}_n), \quad i = 1, ..., s,$$

$$\hat{\mathbf{s}}_i = \underline{\mathbf{u}}^n + \sum_{j=1}^{i-1} \tilde{a}_{ij}\mathbf{U}_j, \quad i = 1, ..., s, \quad (4.31)$$

$$\underline{\mathbf{u}}^{n+1} = \underline{\mathbf{u}}^n + \sum_{i=1}^{s} m_i \mathbf{U}_i.$$

When coding a Rosenbrock method from a given set of coefficients, extra care should be taken, since different authors define the coefficients in slightly different ways, including different scalings by Δt in the definition of the method. Further possible confusion arises from the two different formulations. Here, the coefficients for both formulations will always be given.

As in the case of DIRK methods, the fact that we have a sequence of linear systems can be exploited. In particular, we have a sequence of s linear systems with the same matrix, but varying right-hand side, following in the next time step by a sequence of another s linear systems, with a new matrix that is not arbitrary far away from the last. Approaches for this are described in the next chapter.

One example for a Rosenbrock method is ROS34PW2, which was introduced in [229]. This method has four stages and is of order 3 with an embedded method of order 2. It is a W-method and L-stable. The Butcher array is given in Table B.13. Another example of an L-stable W-method is given by the 6-stage, fourth order method RODASP, which can be seen in Table B.14 and has an embedded method of order 3.

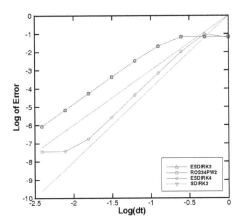

FIGURE 4.4: Order of different time integration methods in the DG case.

A comparison of the order and error of different implicit time integration schemes can be seen in Figure 4.4, where a DG discretization is used for a weakly compressible flow problem. Due to the weak nonlinearity, there is little difference in the errors of Rosenbrock and DIRK schemes. This means that they are competitive with those.

4.6.1 Exponential integrators

A class of time integration methods that has gained increasing interest in the last ten years are exponential integrators. These start from a representation of the right-hand side as a linear function and a nonlinear remainder

$$r(\mathbf{u}) = \mathbf{f}(\mathbf{u}) - \mathbf{f}(\mathbf{u}_n) - \mathbf{f}'(\mathbf{u}_n)(\mathbf{u} - \mathbf{u}_n).$$

With this, the solution can be written as

$$\mathbf{u}(t_n + \Delta t) = \mathbf{u}(t_n) + \Delta t \phi_1(\Delta t \mathbf{f}'(\mathbf{u}_n))\mathbf{f}(\mathbf{u}_n)$$
$$+ \Delta t \int_0^1 e^{\Delta t \mathbf{f}'(\mathbf{u}_n)(1-\Theta)} r(\mathbf{u}(t_n + \Theta \Delta t)) d\Theta,$$

where $\phi_1(z) = (e^z - 1)/z$. Discrete approximations of the exponential of the Jacobian and the nonlinear integral are then needed. For the exponential function, it was discovered that using Krylov subspace approximations results in very efficient schemes [137]. Krylov subspace methods are discussed in detail in section 5.8. Their main property relevant for these methods is that they use Jacobian vector products and scalar products only. For the integral, a quadrature formula is used.

Different approaches are possible and give rise to different classes of exponential integrators. For example, the nonlinear integrand can again be written as a linear function plus a nonlinear remainder. This works well for problems where that remainder is small. Either way, again exponentials of matrices have to be evaluated very efficiently. This gives rise to exponential Runge-Kutta (EPIRK) methods [276], exponential Rosenbrock methods (EXPROS) [137], and others.

Due to using the exact solution for linear problems, these methods are all A-stable. Furthermore, they use only Jacobian vector products and can thus be parallelized easily. Thus, they are interesting methods for high performance computing. A state-of-the-art method is KIOPS [103].

4.7   Adaptive time step size selection

For an explicit method, the CFL condition is typically so strict that when choosing the time step based on it, the time integration error is way below the error tolerance. Thus, we do not need to care about the time step beyond choosing a stable one. This is not the case for an implicit method, which means that a different strategy is needed for the time step selection. Often, a CFL number is stated and the time step computed accordingly. This results in a nonconstant, slowly varying time step. In this approach, we have no idea about how large our time integration error is. Furthermore, if the error is small, we could use larger time steps, which means that the scheme could be more efficient. The same arguments apply to a strategy with a fixed time step.

Therefore, time-adaptive strategies are needed. These consist of two steps: first, the time integration error is estimated and then a controller chooses the new time step based on the error estimate. For DIRK and Rosenbrock methods, the estimate is obtained using embedded schemes of a lower order $\hat{p}$, typically $p - 1$. For IRK methods, it would only be possible to obtain $\hat{p} \ll p$, leading to a bad error estimate and therefore, embedded schemes are not used for these methods. Comparing the local truncation error of both schemes, an estimate of the local error $\hat{\mathbf{l}}$ of the lower order scheme is obtained:

$$\hat{\mathbf{l}} \approx \Delta t_n \sum_{j=1}^{s} (b_j - \hat{b}_j) \mathbf{k}_j, \tag{4.32}$$

respectively, for the efficient formulation (4.31) of the Rosenbrock method

$$\hat{\mathbf{l}} \approx \sum_{j=1}^{s} (m_j - \hat{m}_j) \mathbf{U}_j. \tag{4.33}$$

For BDF methods, the error estimate is obtained by taking two steps with half the time step size, resulting in $\underline{\mathbf{u}}_2$ and using Richardson extrapolation. The local error $\mathbf{l}_2$ after the two half steps for a method of order p is then estimated as

$$\mathbf{l}_2 \approx \frac{\underline{\mathbf{u}}_2 - \mathbf{u}^{n+1}}{2^p - 1}. \tag{4.34}$$

The local error estimate is then used to determine the new step size. To do this, we decide beforehand on a target error tolerance, which we implement using a common fixed resolution test [255]. This means that we define the error tolerance per component by

$$d_i = RTOL|u_i^n| + ATOL. \tag{4.35}$$

Typically, we choose $RTOL = ATOL$, so that there is only one input parameter for this. We then compare (4.35) to the local error estimate by requiring

$$\|\hat{\mathbf{l}}./\mathbf{d}\| \leq 1, \tag{4.36}$$

where the . denotes a pointwise division operator. An important question here is the choice of norm. First of all, the maximum norm guarantees that the tolerance test (4.35) is satisfied in every cell. However, the 2-norm is typically more accurate overall, and since we have to define a tolerance for the subsolver, namely the Newton-GMRES method in some norm, it is natural to use this one, since GMRES uses a 2-norm minimization. Therefore, we will use the latter, since it seems to work well in practice. Nevertheless, a more substantial and less heuristic choice of norm would be desirable for the future.

The next question is how to choose the time step based on the error estimate. Under the assumption that higher-order terms can be neglected in a Taylor expansion of the local error, we have

$$\|\hat{\mathbf{l}}\| \approx \|\Delta t_n^{\hat{p}+1} \hat{\Psi}(t_n, \underline{\mathbf{u}})\| \approx \|\hat{\underline{\mathbf{u}}}^{n+1} - \underline{\mathbf{u}}^{n+1}\|,$$

where $\hat{\Psi}(t_n, \underline{u})$ is the error function of the embedded method. If we additionally assume that $\hat{\Psi}(t, \underline{u})$ is slowly varying, we obtain the classical method (see [122]):

$$\Delta t_{\text{new}} = \Delta t_n \cdot \|\mathbf{1}./\mathbf{d}\|^{-1/k}. \tag{4.37}$$

Here, $k = \hat{p}$ leads to a so-called error per unit step (EPUS) control, whereas $k = \hat{p}+1$ is an error per step (EPS) control. We will use EPS, which often results in more efficient schemes. Formula (4.37) is combined with safety factors to avoid volatile increases or decreases in time step size:

$$\text{if } \|\mathbf{1}./\mathbf{d}\| \geq 1, \qquad \Delta t_{n+1} = \Delta t_n \max(f_{\min}, f_{\text{safety}}\|\mathbf{1}./\mathbf{d}\|^{-1/\hat{p}+1})$$
$$\text{else} \qquad \Delta t_{n+1} = \Delta t_n \min(f_{\max}, f_{\text{safety}}\|\mathbf{1}./\mathbf{d}\|^{-1/\hat{p}+1}).$$

This classical method is called the I-controller in control theory. More elaborate controllers that take the step size history into account are possible, for example the PID controller [253]. Söderlind suggested a controller using methods from signal processing [251]:

The next time step Δt^{n+1} is determined with

$$\Delta t^{n+1} = \hat{\rho}^n \Delta t^n, \tag{4.38}$$

with $\hat{\rho}^n$ given by the smooth limiter

$$\hat{\rho}^n = 1 + \kappa \arctan\left(\frac{\rho^n - 1}{\kappa}\right). \tag{4.39}$$

Choosing $\kappa = 2$ implies that the step size increase is limited to a step doubling, and that a reduction is limited to a factor of approximately 5. In [254], it is noted that the actual value of κ is not crucial. A larger value of κ may allow the step size to increase quickly after the transients have decayed.

ρ^n is determined with the H211PI controller:

$$\rho^n = \|\mathbf{d}^n./\mathbf{l}^n\|^{\beta_1} \|\mathbf{d}^{n-1}./\mathbf{l}^{n-1}\|^{\beta_2} \left(\rho^{n-1}\right)^{-\zeta}, \tag{4.40}$$

where . denotes a pointwise division operator. The coefficients β_1, β_2, and ζ are set to $\beta_1 = \beta_2 = \frac{1}{4\hat{p}}$ and $\zeta = \frac{1}{4}$.

Finally, tolerance scaling and calibration should be used [254]. Tolerance scaling is useful, since, although for the above controller, convergence of the method for $TOL \to 0$ can be proven, the relation between global error and tolerance is typically not such that a decrease of the tolerance by a factor of 10 leads to a decrease of the error by the same factor. Instead, it is observed that

$$\|e\| = \tau \cdot TOL^\alpha,$$

where α is smaller than 1, but does not depend strongly on the problem solved and τ is a proportionality factor. Therefore, an internal rescaling $TOL' =$

$\mathcal{O}(TOL^{1/\alpha})$ should be done, so that the user obtains a behavior where a decrease in TOL leads to a corresponding decrease in error.

Tolerance calibration is used to make sure that at least for a reference problem and a specific tolerance TOL_0, TOL and TOL' are the same. This is achieved via the rescaling

$$TOL' = TOL_0^{(\alpha-1/\alpha)} TOL^{1/\alpha}. \tag{4.41}$$

Through numerical experiments, we suggest the values in Table 4.7 for α for different methods [33].

	SDIRK 3	ESDIRK 3	ESDIRK 4	ROS34PW2
α	0.9	0.9	0.9	0.8

TABLE 4.3: Values for α chosen for different time integration methods

4.8 Operator splittings

Consider an initial value problem with a right-hand side split in a sum of two terms $\underline{\mathbf{f}}_1$ and $\underline{\mathbf{f}}_2$, arising from the discretization of a balance law

$$\frac{d}{dt}\underline{\mathbf{u}}(t) = \underline{\mathbf{f}}_1(t, \underline{\mathbf{u}}(t)) + \underline{\mathbf{f}}_2(t, \underline{\mathbf{u}}(t)), \quad \underline{\mathbf{u}}(t_0) = \underline{\mathbf{u}}^0, \quad t \in [t_0, t_{end}]. \tag{4.42}$$

For example, the splitting could correspond to the x and y dimensions in the Navier-Stokes equations, $\underline{\mathbf{f}}_1$ could be the discretization of the inviscid fluxes, whereas $\underline{\mathbf{f}}_2$ is the discretization of the viscous fluxes. Or $\underline{\mathbf{f}}_1$ could be the sum of viscous and inviscid fluxes, whereas $\underline{\mathbf{f}}_2$ arises from the discretization of a source term like the gravitational force.

The terms $\underline{\mathbf{f}}_1$ and $\underline{\mathbf{f}}_2$ interact and the most accurate way to treat these terms is to respect this interaction in the spatial discretization. This was discussed in section 3.5. However, it is sometimes easier to treat these terms separately. This is called an *operator splitting* or *fractional step* method. These split the solution process into the solution of two ordinary differential equations, which is useful if for both equations, well known methods are available. A first-order approximation (for smooth solutions $\mathbf{u}$) to this problem is given by the simple Godunov splitting [108], also called Lie splitting:

1. Solve $\frac{d}{dt}\mathbf{u} + \underline{\mathbf{f}}_1(\mathbf{u}) = \mathbf{0}$ with time step Δt and initial data $\mathbf{u}^n$ to obtain intermediate data $\mathbf{u}^*$.

2. Solve $\frac{d}{dt}\mathbf{u} + \underline{\mathbf{f}}_2(\mathbf{u}) = \mathbf{0}$ with the same time step, but initial data $\mathbf{u}^*$ to obtain $\mathbf{u}^{n+1}$.

Here, we require each "solve" to be at least first-order accurate.

To increase the accuracy and obtain a scheme of second order for smooth solutions, we have to use a slightly more sophisticated method, for example the Strang splitting [262], where again the subproblem solvers have to be at least of first order:

1. Solve $\frac{d}{dt}\mathbf{u} + \underline{\mathbf{f}}_1(\mathbf{u}) = \mathbf{0}$ with time-step $\Delta t/2$.

2. Solve $\frac{d}{dt}\mathbf{u} + \underline{\mathbf{f}}_2(\mathbf{u}) = \mathbf{0}$ with time-step Δt.

3. Solve $\frac{d}{dt}\mathbf{u} + \underline{\mathbf{f}}_1(\mathbf{u}) = \mathbf{0}$ with time-step $\Delta t/2$.

As before, in each step, the intermediate result obtained in the last step is used as initial data. Since the problem with $\underline{\mathbf{f}}_1$ is solved twice in this operator splitting, $\underline{\mathbf{f}}_1$ should be chosen such that it is easier to solve than the problem with $\underline{\mathbf{f}}_2$. Splittings of order greater than 2 can only be defined under severe assumptions on the operators [43].

The role of $\underline{\mathbf{f}}_1$ and $\underline{\mathbf{f}}_2$ can of course be exchanged; however, in general they do not commute. This becomes obvious when considering a local heat source. Increasing the heat first and then applying the convective flux leads to a different result compared to doing the convective step first and then increasing the heat locally. For this reason, special care has to be taken in choosing the numerical boundary conditions for the partial differential equation. Otherwise, unphysical effects can be introduced into the solution.

The "solves" in each step correspond to a time integration procedure, which has to be chosen of the appropriate order. For the source term, simple time integration procedures can be chosen, possibly leading to more efficient overall schemes than when incorporating the source term into the computation of the fluxes. For the Navier-Stokes equations, an important idea is to use an operator splitting where an implicit time integration method is used for the diffusive fluxes and an explicit method is used for the convective parts, since the severe restriction CFL condition is typically less severe than the DFL condition. This is sometimes referred to as an IMEX scheme for implicit-explicit, but care has to be taken since this term is also used for schemes where an explicit or implicit scheme is used depending on the part of the spatial domain considered. These ideas are similar to the additive RK methods presented in 4.5.4.

Tang and Teng [271] proved for multidimensional scalar balance laws that if the exact solution operator is used for both subproblems, the described schemes converge to the weak entropy solution and furthermore that the L^1 convergence rate of both fractional step methods is not worse than $1/2$. This convergence rate is actually optimal, if a monotone scheme is used for the homogenous conservation law in combination with the forward Euler method for the time integration. Langseth, Tveito, and Winther [171] proved for scalar one-dimensional balance laws that the L^1 convergence rate of the Godunov splitting (again using the exact solution operators) is linear and showed corresponding numerical examples, even for systems of equations. A better convergence rate than linear for nonsmooth solutions is not possible, as Crandall and Majda proved already in 1980 [66].

The L^1 order does not tell the whole story. Using the Strang or Godunov splitting combined with a higher-order method in space and a second-order time integration does improve the solution compared with first-order schemes and is therefore appropriate for the computation of unsteady flows. This is, for example, suggested by LeVeque [175].

4.9 Alternatives to the method of lines

4.9.1 Space-time methods

So far, we have looked at the method of lines only. An alternative to discretizing in space first and then in time is to discretize in space and time simultaneously. This approach is unusual in the finite volume context, but followed for example in the ADER method [275]. For DG methods, this is slightly more common, for example in the space-time DG of Klaij et al. [157] or the space-time expansion (STE) DG of Lörcher et al. [180], which allow for a local time stepping via a predictor-corrector scheme to increase efficiency for unsteady flows. There and in other approaches, the time integration over the interval $[t_n, t_{n+1}]$ is embedded in the overall DG formulation. This was exploited in [93] to prove that a space-time DG-SEM method, using a specific flux function in space and an upwind flux in time, is a linearly stable SBP operator and satisfies a nonlinear entropy stability property for hyperbolic conservation laws. DG-SEM in time corresponds to fully implicit RK methods, specifically Gauß-Lobatto IIIc methods.

4.9.2 Local time stepping Predictor-Corrector-DG

As an example of an alternative method, we will now explain the Predictor-Corrector-DG method and the local time stepping used in more detail. Starting point is the evolution equation (3.32) for the cell i, integrated over the time interval $[t_n, t_{n+1}]$:

$$\mathbf{u}_i^{n+1} = \mathbf{u}_i^n - \int_{t_n}^{t_{n+1}} \underbrace{\sum_{i=1}^{nFaces} \mathbf{M}_i^S \mathbf{g_i}}_{\mathbf{R}_V(p_i)} - \underbrace{\sum_{k=1}^{d} \mathbf{S}_k \mathbf{f}_k}_{\mathbf{R}_S(p_i, p_j)} dt.$$

The time integral is approximated using Gaussian quadrature. This raises the question of how to obtain the values at future times. To this end, the integral is split into the volume term defined by $\mathbf{R}_V$ that needs information from the cell i only, and the surface term $\mathbf{R}_S$ that requires information from neighboring cells. Then, the use of cellwise predictor polynomials $\mathbf{p}_i(t)$ in time is suggested. Once this is given, the update can be computed by

$$\mathbf{u}_i^{n+1} = \mathbf{u}_i^n - \int_{t_n}^{t_{n+1}} \mathbf{R}_V(\mathbf{p}_i) - \mathbf{R}_S(\mathbf{p}_i, \mathbf{p}_j) dt. \tag{4.43}$$

Several methods to obtain these predictor polynomials have been suggested. The first idea was to use a space time expansion by the Cauchy-Kowalewskaja procedure [180], which leads to a rather costly and cumbersome scheme. However, in [98], the use of continuous extension Runge-Kutta (CERK) methods is suggested instead. This is a type of RK methods that allow to obtain approximations to the solution not only at t_{n+1}, but at any value $t \in [t_n, t_{n+1}]$ [218]. The only difference to the dense output formulas (4.24) and (4.25) mentioned earlier is that the latter are designed for particular RK methods, whereas the CERK methods are full RK schemes in their own right. Coefficients of an explicit four-stage CERK method with its continuous extension can be found in the appendix Tables B.15 and B.16.

The CERK method is then used to integrate the initial value problem

$$\frac{d}{dt}\mathbf{u}_i(t) = \mathbf{R}_V(\mathbf{u}_i), \quad \mathbf{u}_i(t_n) = \mathbf{u}_i^n, \tag{4.44}$$

in every cell i, resulting in stage derivatives $\mathbf{k}_j$. The values at the Gauss points are obtained via the CERK polynomial,

$$\mathbf{p}(t) = \sum_{k=0}^{p} \mathbf{q}_k t^k$$

which is of order p corresponding to the order of the CERK method minus 1 and has the coefficients

$$\mathbf{q}_k = \frac{1}{\Delta t_n^k} \sum_{j=1}^{s} b_{kj} \mathbf{k}_j,$$

where the coefficients b_{kj} can be found in the appendix table B.16.

The predictor method needs to be of order $k-1$, if order k is sought for the complete time integration. If a global time step is employed, the method described so far is already applicable. However, a crucial property of the scheme is that a local time stepping procedure can be used.

A cell i can be advanced in time, if the necessary information in the neighboring cells is already there; thus if the local time at time level n is not larger than the time level in the neighboring cells:

$$t_i^{n+1} \leq \min\left\{t_j^{n+1}\right\} \forall j \in N(i). \tag{4.45}$$

This is illustrated in Figure 4.5. There, all cells are synchronized at time level t_n, then in each cell a predictor polynomial is computed using the CERK method, which is valid for the duration of the local time step Δt_{n_i}. However, for the completion of a time step in both cell $i-1$ and $i+1$, boundary data is missing. However, cell i fulfills the evolve condition and thus, the time step there can be completed using the predictor polynomials p_{i-1} and p_{i+1}. Then, the predictor polynomial in cell i for the next local time step is computed. Now, cell $i-1$ fulfills the evolve condition and after the completion of that time step, the second time step in cell i can be computed. Note that while this example looks sequential, on a large grid, a number of cells can be advanced in time in parallel, making the scheme attractive on modern architectures.

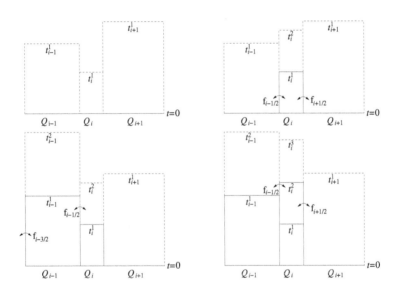

FIGURE 4.5: Sequence of steps 1–4 of a computation with three different elements and local time-stepping. (Credit: M. Ferch.)

Finally, to make the scheme conservative, care needs to be taken when a cell is advanced in time, whose neighbor has been partially advanced. Then, the appropriate time integral is split into two parts, which are computed separately:

$$\int_{t^n}^{t^{n+1}} \ldots dt = \int_{t^n}^{t^1} \ldots dt + \int_{t^1}^{t^2} \ldots dt + \cdots + \int_{t^{K-2}}^{t^{K-2}} \ldots dt + \int_{t^{K-1}}^{t^{n+1}} \ldots dt,$$

we split the interval $\left[t_1^n, t_1^{n+1}\right]$ into the intervals $\left[t_1^n, t_2^{n+1}\right]$ and $\left[t_2^{n+1}, t_1^{n+1}\right]$ which yields

$$\int_{t_1^n}^{t_1^{n+1}} \mathbf{R}_S\left(\tilde{p}_1, \tilde{p}_2\right) dt = \int_{t_1^n}^{t_2^{n+1}} \mathbf{R}_S\left(\tilde{p}_1^n, \tilde{p}_2^n\right) dt + \int_{t_2^{n+1}}^{t_1^{n+1}} \mathbf{R}_S\left(\tilde{p}_1^n, \tilde{p}_2^{n+1}\right) dt.$$

4.10   Parallelization in time

With hardware getting massively parallel, the question of time integration schemes that are parallel by themselves has and will continue to become more important. Since time integration is inherently sequential, the idea of parallelizing it is counterintuitive. This is confirmed by the methods available: Parallelization in space or running multiple instances of the same code at the same time will always be more efficient. However, for any given problem, there is a limit to parallelization in space, since at some point, communication becomes dominant. Thus, for most turbulent flow problem, parallelization in time is not on the agenda and will not be for the foreseeable future. However, for laminar or inviscid flows, this has to be considered as an option. A survey of the approaches so far can be found in [94].

A first such method has been presented in section 4.5.2: Fully implicit RK methods, where all stages are solved in parallel. This type of parallelization is very limited by the number of stages. The alternative is to choose a time slab with k time steps and solve all these steps simultaneously in parallel. The first idea in this regard was parareal [96] and similar ones are spectral deferred correction [80, 203] and integral deferred correction [60]. These have been combined with multigrid ideas in PFASST [85, 46] and MGRIT [88].

An alternative angle comes from space-time methods. There, an implicit discretization for a whole time slab of multiple time steps is written down as a huge equation system. This can then be solved in parallel using specially designed equation solvers. As an example, in [184], Mani and Mavriplis do this for a space-time-adaptive FV discretization.

Chapter 5

Solving equation systems

The application of an implicit scheme for the Navier-Stokes equations leads to a nonlinear or, in the case of Rosenbrock methods, linear system of equations. To solve systems of this form, different methods are used. As mentioned in the Introduction, the question of efficiency of an implicit scheme will be decided by the solver for the equation systems. Therefore, this chapter is the longest of this book.

The outline is as follows: We will first describe properties of the systems at hand and general paradigms for iterative solvers. Then we will discuss methods for nonlinear systems, namely *fixed point methods*, *multigrid methods*, and different variants of *Newton's method*. In particular, we make the point that dual time stepping multigrid as currently applied in industry can be vastly improved and that inexact Newton methods are an important alternative. We will then discuss Krylov subspace methods for the solution of linear systems and finally revisit Newton methods in the form of the easy-to-implement Jacobian-free Newton-Krylov methods.

5.1 The nonlinear systems

For DIRK or BDF methods, we obtain systems of the form (see (4.21), (4.23), and (4.12))

$$\underline{u} = \underline{\tilde{u}} + \alpha \Delta t \hat{\underline{f}}(\underline{u}), \tag{5.1}$$

where $\underline{u} \in \mathbb{R}^m$ is the vector of unknowns, α a parameter and $\underline{\tilde{u}}$ is a given vector. As before, the underbar denotes a vector of all conservative variables from all cells. Finally, the function $\hat{\underline{f}}(\underline{u})$ consists of everything else coming from the spatial and temporal discretization. In the case of an autonomous ODE, $\hat{\underline{f}}(\underline{u})$ just denotes an evaluation of the function $\underline{f}(\underline{u})$ representing the spatial discretization on the whole grid.

If (5.1) has a unique solution, we call it $\underline{u}^*$. For nonlinear systems, there are two important formulations that are used depending on the context. The first one is the fixed point form

$$\underline{u} = \underline{g}(\underline{u}) \tag{5.2}$$

and the second one is the root form

$$\underline{\mathbf{F}}(\underline{\mathbf{u}}) = \underline{\mathbf{0}}.\tag{5.3}$$

Equation (5.1) is in fixed point form, and one possible root form would be

$$\underline{\mathbf{u}} - \tilde{\underline{\mathbf{u}}} - \alpha \Delta t \hat{\underline{\mathbf{f}}}(\underline{\mathbf{u}}) = \underline{\mathbf{0}}.$$

Obviously, neither form is unique for a given nonlinear equation.

For the right-hand side functions arising in CFD, formulas for exact solutions of (5.1) do not exist, which is why iterative methods are needed. These produce a sequence of iterates $\{\underline{\mathbf{u}}^{(k)}\}$, hopefully converging to the solution $\underline{\mathbf{u}}^*$. There exists a plethora of schemes to solve multidimensional nonlinear systems, for example fixed point methods, Newton's method and its variants or multigrid methods. All of these are employed in the CFD context, with fixed point methods being used due to high robustness, which means global and provable convergence, whereas the other two are much faster. We will examine those in detail in this chapter. First, we will discuss some properties of the nonlinear equation system.

Important mathematical questions are whether there exist solutions of a given equation and if these are unique. As mentioned in chapter 2, both of these have not been answered for the continuous equations. In fact, it is known that the steady Euler equations allow nonunique solutions, and for the Navier-Stokes equations, existence and uniqueness results for general data are an open question. With regard to the discrete equations we are faced with, we know that if the right-hand side is Lipschitz continuous, then the later presented Banach's fixed point theorem tells us that for a sufficiently small Δt, equation (5.1) has a unique solution. For larger time step sizes or right-hand sides that are not Lipschitz continuous, no results are known.

Regarding the properties of the discrete system, an important role is played by numerical flux functions, of which we required that they are consistent, which in particular implied Lipschitz continuity, but not differentiability! However, a look at typical numerical flux functions tells us that they are differentiable except for a finite number of points. In the finite volume context, we have to look at the reconstruction and the limiters as well. The first one is based on linear least squares problems, which means that the reconstructed values depend differentiably on the data. On the other hand, the latter always contain minimum or maximum functions and are therefore differentiable except at a finite number of points and Lipschitz continuous otherwise. Regarding DG methods, we know that the cellwise approximations are even differentiable and the only problem is posed by the numerical flux functions. Finally, we have to consider turbulence models and these are again in general only Lipschitz continuous and piecewise differentiable.

All these components are then combined using sums and products, which means that the resulting function is globally Lipschitz continuous and except for a finite number of points, also differentiable.

5.2 The linear systems

In the case of a Rosenbrock method (4.30) or if the nonlinear systems are solved using Newton's method, we obtain linear systems of the form

$$\mathbf{A}\mathbf{x} = \mathbf{b} \tag{5.4}$$

with $\mathbf{x}, \mathbf{b} \in \mathbb{R}^m$ and $\mathbf{A} \in \mathbb{R}^{m \times m}$ with

$$\mathbf{A} = \left. \frac{\partial \mathbf{F}(\underline{\mathbf{u}})}{\partial \underline{\mathbf{u}}} \right|_{\underline{\mathbf{u}}} = \left(\mathbf{I} - \alpha \Delta t \frac{\partial \hat{\mathbf{f}}(\underline{\mathbf{u}})}{\partial \underline{\mathbf{u}}} \right) \Bigg|_{\underline{\mathbf{u}}} . \tag{5.5}$$

The values of the Jacobian and thus the cost of computing it depend on the set of unknowns chosen, e.g., primitive or conservative. Often, the formulas are simpler when using primitive variables, leading to a small but noticeable speedup.

The matrix $\mathbf{A}$ in (5.5) has the property of being sparse, ill conditioned, and unsymmetric (though it has a certain block symmetry). We can also deduce that the matrix is close to the identity matrix for small time steps and that thus the linear equation systems become the harder to solve, the bigger the time step is.

As for the block structure of $\mathbf{A}$, this depends on the space discretization method. The size of the blocks corresponds to the number of unknowns in a cell. For a finite volume scheme, the size is $d + 2$ and the blocks are dense, whereas for a modal DG scheme, the size increases to $(p+d)!/(p!d!)$ with again dense blocks. However, while the block size for the DG-SEM is even greater for the same order with $(p + 1)^d$ unknowns, the blocks are suddenly sparse with $(d + 1)dp^d$ nonzero entries for the case of the Euler equations. This can be immediately seen from (3.37) and (3.38) and is visualized in Figure 5.1.

In the case of the Navier-Stokes equations, the computation of gradients leads to a less sparse block. However, due to using the dGRP-flux (3.45) or similar approaches, we still have some sparsity, as shown in Figure 5.2. Regarding the offdiagonal blocks, we again have to take into account the gradients of the points on the boundary in neighboring cells.

Since the number of unknowns is rather large and can be several millions in the case of 3D-Navier-Stokes computations, it is of utmost importance to use the structure of this equation system to obtain a viable method. Thereby, both storage and computing time need to be considered. For example, it is impossible to store a dense Jacobian matrix and therefore, the sparsity of the Jacobian must be exploited. This narrows the choice down to sparse direct methods, splitting methods, Krylov subspace methods, and linear multigrid methods.

Splitting methods like Jacobi or Gauss-Seidel can exploit the sparsity, but they are only linearly convergent with a constant near 1, which makes them

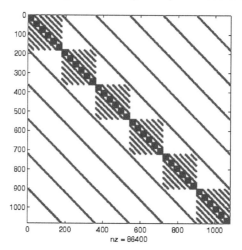

FIGURE 5.1: Sparsity pattern of a diagonal block for the Euler equations. (Credit: M. Ferch.)

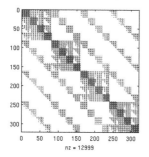 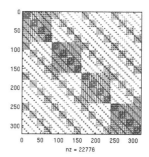

FIGURE 5.2: Sparsity pattern of a diagonal (left) and offdiagonal (right) for the Navier-Stokes equations. (Credit: M. Ferch.)

too slow. However, they will play a role as preconditioners and smoothers for other methods. As for linear multigrid methods, there is no theory that tells us what smooth components are for the Euler- or Navier-Stokes equations, and therefore, no fast linear multigrid solver has been found. Regarding sparse direct methods, there has been significant progress during the past decades and robust solver packages like PARDISO, SuperLU, or UMFPACK have been developed. At the moment, these are still slower than Krylov subspace methods for large systems, but might be an option in the future [126]. The major choice left are therefore so-called Krylov subspace methods.

5.3 Rate of convergence and error

Since for both nonlinear and linear systems iterative methods play an important role, it helps to have properties that can be used to compare these schemes. One is the notion of the rate of convergence.

Definition 7 (Rate of convergence) *A method with iterates* $\mathbf{x}^{(k)} \in \mathbb{R}^m$, $k \in \mathbb{N}$, *which converges to* $\mathbf{x}^*$ *is called*

- *linearly convergent to* $\mathbf{x}^*$, *if* $\|\mathbf{x}^{(k+1)} - \mathbf{x}^*\| \leq C\|\mathbf{x}^{(k)} - \mathbf{x}^*\|$, $0 < C < 1$,

- *superlinearly convergent of order p to* $\mathbf{x}^*$, *if* $\|\mathbf{x}^{(k+1)} - \mathbf{x}^*\| \leq C\|\mathbf{x}^{(k)} - \mathbf{x}^*\|^p$ *with* $p > 1$, $C > 0$,

- *superlinearly convergent to* $\mathbf{x}^*$, *if* $\lim_{k \to \infty} \frac{\|\mathbf{x}^{(k+1)} - \mathbf{x}^*\|}{\|\mathbf{x}^{(k)} - \mathbf{x}^*\|} = 0$,

- *quadratically convergent to* $\mathbf{x}^*$, *if* $\|\mathbf{x}^{(k+1)} - \mathbf{x}^*\| \leq C\|\mathbf{x}^{(k)} - \mathbf{x}^*\|^2$ *with* $C > 0$.

Note that superlinear convergence of order p implies superlinear convergence, but not the other way round.

Using the error for a given vector $\underline{\mathbf{u}}$

$$\underline{\mathbf{e}} := \underline{\mathbf{u}} - \underline{\mathbf{u}}^* \tag{5.6}$$

and using the shorthand notation for the error in the kth iteration

$$\underline{\mathbf{e}}_k := \underline{\mathbf{u}}^{(k)} - \underline{\mathbf{u}}^*, \tag{5.7}$$

we can write these notions in quite compact form.

5.4 Termination criteria

For iterative schemes, it is important to have a termination criterion that avoids costly oversolving, i.e., beyond the accuracy needed. Furthermore, a norm has to be chosen. Often the 1-, 2-, or ∞-norm are used with a fixed tolerance for each system, to be defined in the parameter file. While easy, this neglects that we do not solve these nonlinear systems for themselves, but inside a time-adaptive implicit time integration procedure. This means that the nonlinear systems have to be solved to a degree that guarantees not to interfere with the error estimator in the time integration, such as not to lead to unpredictable behavior there. In [254], it is demonstrated on the DASSL code, how a bad choice of the tolerance in the nonlinear solver can lead to problems in the adaptive time integrator. The way to circumvent this is to choose the tolerances in the nonlinear solver based on the tolerances in the time integrator (see (4.35)) and furthermore, to measure convergence in the same norm as for the error estimator. Thus, we will always assume that a tolerance TOL and a norm $\|\cdot\|$ will be given from the outside solver.

We are now faced with the same problem as for the time integration: We do not know the error and have to estimate it instead. In the absence of any other information about the system, typically one of the following two quantities is used to estimate the error: The residual $\|\mathbf{F}(\mathbf{u}^{(k)})\|$ given by (5.3) or the difference of two subsequent iterates $\|\mathbf{u}^{(k+1)} - \mathbf{u}^{(k)}\|$. Obviously, both converge to zero if the sequence converges to $\mathbf{u}^*$, but conversely, it is not true that when one of these is small, we can be sure that the sequence converges to the solution or that $\mathbf{u}^{(k)}$ is close to $\mathbf{u}^*$. Nevertheless, the term "convergence criteria" is often used instead of termination criteria and it is common to say that a solver has converged to denote that the termination criterion is satisfied.

For a differentiable function $\mathbf{F}$ we have the following relation between relative residual and relative error:

$$\frac{\|\mathbf{e}\|}{4\|\mathbf{e}_0\|\kappa(\mathbf{F}'(\mathbf{u}^*))} \leq \frac{\|\mathbf{F}(\mathbf{u})\|}{\|\mathbf{F}(\mathbf{u}^{(0)})\|} \leq \frac{4\kappa(\mathbf{F}'(\mathbf{u}^*))\|\mathbf{e}\|}{\|\mathbf{e}_0\|},$$

where κ denotes the condition number in the 2-norm. The inequalities mean that if the Jacobian of the function in the solution is well conditioned, the residual is a good estimator for the error. If this is not the case, then we do not know. For the other criteria, we have that

$$\mathbf{e}_{k+1} = \mathbf{e}_k + \mathbf{u}^{(k+1)} - \mathbf{u}^{(k)}$$

and thus for a method that is convergent of order p we have

$$\|\mathbf{e}_k\| = \|\mathbf{u}^{(k+1)} - \mathbf{u}^{(k)}\| + \mathcal{O}(\|\mathbf{e}_k\|^p)$$

near the solution.

As termination criteria, we always use relative ones. For the residual-based indicator, we obtain

$$\|\mathbf{F}(\mathbf{u}^{k+1})\| \leq TOL \cdot \|\mathbf{F}(\mathbf{u}^{(0)})\| \tag{5.8}$$

and for the solution-based indicator, the test becomes

$$\|\mathbf{u}^{(k+1)} - \mathbf{u}^{(k)}\| \leq TOL\|\mathbf{u}^{(0)}\|. \tag{5.9}$$

For the solution of steady problems when using a time integration scheme to compute the steady state, it is typically sufficient to do only a very small number of iterations of the nonlinear solver. This is not the case for unsteady problems, where it is important that the termination criterion is reached for not interfering with the outer time integration scheme.

Now, it may happen that the time step is chosen so large by the adaptive time step selector, that the nonlinear solver does not converge. Therefore, it is useful to add another feedback loop to this: Set a maximal number of iterations for the nonlinear solver. If the iteration has not passed the termination criterion by then, repeat the time step, but divide the step size by 2. This is summarized in chapter 7.

5.5 Fixed point methods

As mentioned before, we speak of a fixed point equation, if it is in the form

$$\underline{\mathbf{g}}(\underline{\mathbf{u}}) = \underline{\mathbf{u}}. \tag{5.10}$$

Note that (5.1) is in fixed point form. A method to solve such equations is the fixed point iteration

$$\underline{\mathbf{u}}^{(k+1)} = \underline{\mathbf{g}}(\underline{\mathbf{u}}^{(k)}). \tag{5.11}$$

A very useful and important theorem to determine convergence of this type of methods is Banach's fixed point theorem:

Theorem 5 *Let X be a Banach space with norm $\|\cdot\|$. Let $D \subset X$ be a closed set and $\underline{\mathbf{g}}$ be a contraction on D, meaning that $\underline{\mathbf{g}}$ is Lipschitz continuous on D with Lipschitz constant $L_g < 1$. Then the fixed point equation (5.10) has a unique solution $\underline{\mathbf{u}}^*$ and the iteration (5.11) converges linearly to this solution if $\underline{\mathbf{u}}^{(0)} \in D$.*

When applying this theorem to equation (5.1), the important function is $\Delta t \underline{\hat{\mathbf{f}}}(\underline{\mathbf{u}})$, which has a Lipschitz constant of $\Delta t L_f$, where L_f is the Lipschitz constant of $\underline{\hat{\mathbf{f}}}$. This implies that for convergence of the fixed point iteration, $\Delta t L_f < 1$ has to be satisfied, which puts an additional constraint on the time step. If the problem is very stiff, this leads to an unacceptably strict constraint, even for A-stable methods.

5.5.1 Stationary linear methods

A particular type of fixed point methods for the solution of linear systems $\mathbf{A}\mathbf{x} = \mathbf{b}$, $\mathbf{A} \in \mathbb{R}^{m \times m}$ are stationary linear methods:

$$\mathbf{x}^{(k+1)} = (\mathbf{I} - \mathbf{N}^{-1}\mathbf{A})\mathbf{x}^{(k)} + \mathbf{N}^{-1}\mathbf{b}, \tag{5.12}$$

where $\mathbf{N} \in \mathbb{R}^{m \times m}$ is a nonsingular matrix. These methods, while not fast for the problems at hand, form basic building blocks of iterative solvers and are useful as subcomponents of other later discussed methods, for example as smoothers in multigrid or preconditioners in Krylov subspace methods. Their name comes from writing the system matrix in a split form

$$\mathbf{A} = (\mathbf{A} - \mathbf{N}) + \mathbf{N}$$

and then defining a fixed point equation, which is equivalent to $\mathbf{A}\mathbf{x} = \mathbf{b}$ by

$$\mathbf{x} = (\mathbf{I} - \mathbf{N}^{-1}\mathbf{A})\mathbf{x} + \mathbf{N}^{-1}\mathbf{b}. \tag{5.13}$$

Thus, in case of convergence, the limit of the iteration (5.12) is both a fixed (stationary) point of (5.13) and a solution of $\mathbf{A}\mathbf{x} = \mathbf{b}$.

The matrix $\mathbf{M} := \mathbf{I} - \mathbf{N}^{-1}\mathbf{A}$ is called iteration matrix and determines whether the iteration converges. This is formalized in Theorem 6.

Theorem 6 *Let* $\mathbf{x}^{(k+1)} = \mathbf{M}\mathbf{x}^{(k)} + \mathbf{N}^{-1}\mathbf{b}$ *be a linear iterative scheme. Then the scheme is globally convergent for any* $b \in \mathbb{R}^m$ *if and only if* $\rho(\mathbf{M}) < 1$.

This can be proved using Banach's fixed point theorem and the fact that for any $\epsilon > 0$, there is a matrix norm, such that for any matrix $\mathbf{M} \in \mathbb{R}^{m \times m}$, it holds that $\|\mathbf{M}\| \leq \rho(\mathbf{M}) + \epsilon$.

Thus, if $\|\mathbf{I} - \mathbf{N}^{-1}\mathbf{A}\| < 1$ or more general if $\rho(\mathbf{I} - \mathbf{N}^{-1}\mathbf{A}) < 1$, the methods converge. In the case of block matrices, this was considered by Varga [290], and in the context of computational fluid dynamics, by Dwight [81].

As a termination criterion, instead of testing if equation (5.13) is satisfied, the relative criterion (5.8) should be used, based on the linear residual $\mathbf{r} = \mathbf{b} - \mathbf{A}\mathbf{x}^{(k)}$.

Several specific choices for the matrix $\mathbf{N}$ correspond to well-known methods. The simplest option is $\mathbf{N} = \mathbf{I}$, leading to the *Richardson iteration*

$$\mathbf{x}^{(x+1)} = (\mathbf{I} - \mathbf{A})\mathbf{x}^{(k)} + \mathbf{b}. \tag{5.14}$$

The idea of the *Gauß-Seidel method* is to start with the first equation and solve for the first unknown, given the current iterate. The first component of the first iterate is then overwritten and the method proceeds to the next equation, until each equation has been visited. Thus, the method is inherently sequential and furthermore, if the system matrix is triangular, it is a direct solver. In matrix notation, it can be written as

$$\mathbf{x}^{(x+1)} = -(\mathbf{L} + \mathbf{D})^{-1}\mathbf{U}\mathbf{x}^{(k)} + (\mathbf{L} + \mathbf{D})^{-1}\mathbf{b}, \tag{5.15}$$

where $\mathbf{L}$ is the strict lower left part of $\mathbf{A}$, $\mathbf{D}$ the diagonal, and $\mathbf{U}$ the strict upper right part. Thus, $\mathbf{N} = \mathbf{L} + \mathbf{D}$. The action of the inverses can be computed using a forward solve.

If, following the forward sweep of the Gauß-Seidel method, a backward sweep using the same method with a reverse numbering of the unknowns is performed, the *symmetric Gauß-Seidel method* (SGS) is obtained. This can be written using

$$\mathbf{N}_{SGS} = (\mathbf{D} + \mathbf{L})\mathbf{D}^{-1}(\mathbf{D} + \mathbf{U}), \tag{5.16}$$

leading to

$$\mathbf{x}^{(x+1)} = x^{(k)} - (\mathbf{D} + \mathbf{U})^{-1}\mathbf{D}(\mathbf{D} + \mathbf{L})^{-1}(\mathbf{A}\mathbf{x}^{(k)} - \mathbf{b}), \tag{5.17}$$

where the application of the inverses is again implemented using forward and backward solves.

The convergence behavior of both GS and SGS depends on the ordering of the unknowns. Therefore, if this method is used in some way, we suggest to reorder the unknowns appropriately after grid generation or adaption. A particular strategy is to use physical reordering as suggested in [199]. There, planes are built orthogonal to the direction of the inflow and then the unknowns in the first plane are numbered first, then the second plane, and so on. This significantly improves the performance of SGS, as well as of the later described ILU preconditioner.

Furthermore, there is the *Jacobi method*, which is inherently parallel and is obtained by choosing

$$\mathbf{N}_J = \mathbf{D}. \tag{5.18}$$

The idea behind the method is to solve all equations simultaneously for the unknown on the diagonal, given the current values for the other unknowns. In matrix notation, it can be written as

$$\mathbf{x}^{(x+1)} = -\mathbf{D}^{-1}(\mathbf{L} + \mathbf{U})\mathbf{x}^{(k)} + \mathbf{D}^{-1}\mathbf{b}. \tag{5.19}$$

Finally, all these methods exist in a relaxation form, meaning that a relaxation parameter ω is added, which can lead to improved convergence speed. The corresponding variant of the Richardson iteration is

$$\mathbf{x}^{(x+1)} = (\mathbf{I} - \omega\mathbf{A})\mathbf{x}^{(k)} + \omega\mathbf{b} \tag{5.20}$$

For SGS, this is known as the *symmetric overrelaxation method* (SSOR) and is given by

$$\mathbf{N}_{\text{SSOR}} = \frac{1}{\omega(2-\omega)}\mathbf{N}_{\text{SGS}} = \frac{1}{\omega(2-\omega)}(\mathbf{D}+\mathbf{U})^{-1}\mathbf{D}(\mathbf{D}+\mathbf{L})^{-1}. \tag{5.21}$$

All these methods can be extended trivially to block matrices.

5.5.2 Nonlinear variants of stationary methods

The Jacobi and Gauß-Seidel iteration can be extended to nonlinear equations [234]. To this end, recall the original idea of the method. Thus, the Gauß-Seidel process starts with the first nonlinear equation and, given the current iterate $\underline{\mathbf{u}}^{(k)}$, solves for the first unknown $u_1^{(k)}$. The first component of the first iterate is then overwritten and the method proceeds to the next equation, until each equation has been visited. Here, solve means an inner iterative process for the solution of nonlinear systems, for example, Newton's method. Sometimes, the inner solve is added to the name, for example, Gauß-Seidel-Newton process. We thus obtain the method

- For $k = 1, \ldots$ do

  - For $i = 1, \ldots, m$ do:

    Solve $f_i(u_1^{(k+1)}, \ldots, u_{i-1}^{(k+1)}, u_i^{(k+1)}, u_{i+1}^{(k)}, \ldots u_n^{(k)}) = 0$ for $u_i^{(k+1)}$.
    $$(5.22)$$

The symmetric Gauß-Seidel and the Jacobi method can be extended in a corresponding way; for example, the Jacobi process can be written as

- For $k = 1, \ldots$ do

  - For $i = 1, \ldots, n$ do:

    Solve $f_i(u_1^{(k)}, \ldots, u_{i-1}^{(k)}, u_i^{(k+1)}, u_{i+1}^{(k)}, \ldots u_n^{(k)}) = 0$ for $u_i^{(k+1)}$.
    $$(5.23)$$

Again, for all of these processes, relaxation can be used to improve convergence. Furthermore, the extension to a block form, where instead of scalar equations, nonlinear systems have to be solved in each step, is straight forward.

Regarding convergence, we can again use Banach's fixed point theorem to deduce under the assumption of exact solves, that this will be linear. More precise, we define a mapping $\mathbf{g}$ for the Gauß-Seidel case via the formula

$$g_i(\mathbf{x}, \mathbf{y}) := f_i(x_1, \ldots, x_i, y_{i+1}, \ldots, y_n)$$

for its ith component. The iteration is then implicitly defined by the equation

$$\mathbf{g}(\underline{\mathbf{u}}^{(k+1)}, \underline{\mathbf{u}}^{(k)}) = \mathbf{0}.$$

Then we are in a fixed point of the original equation if $\mathbf{x} = \mathbf{y}$, corresponding to $\underline{\mathbf{u}}^{(k+1)} = \underline{\mathbf{u}}^{(k)}$. It then follows using the implicit function theorem that the process converges in a neighborhood of the solution $\underline{\mathbf{u}}^*$ if

$$\rho(-\partial_1\mathbf{g}(\underline{\mathbf{u}}^*, \underline{\mathbf{u}}^*)^{-1}\partial_2\mathbf{g}(\underline{\mathbf{u}}^*, \underline{\mathbf{u}}^*)) < 1,$$

where the partial derivatives are taken with respect to $\mathbf{x}$ and $\mathbf{y}$ [234, p. 31]. This corresponds to the condition on $\mathbf{I} - \mathbf{B}^{-1}\mathbf{A}$ in the linear case.

5.6 Multigrid methods

A class of methods that has been developed particularly for equation systems arising from discretized partial differential equations are multigrid methods [120, 281, 302]. These need to be tailored to the specific PDE. If designed properly, multigrid schemes are linearly convergent and furthermore, the so-called textbook multigrid efficiency has been demonstrated or even proved for large classes of partial differential equations, in particular elliptic ones. This means that the convergence rate is independent of the mesh width and that only a few steps are necessary to compute the solution. Multigrid methods are the standard method used in industry codes, even though textbook multigrid efficiency has not been achieved for the Navier-Stokes equations. The theory is well developed for elliptic problems, but still lacking for other types.

The idea is to divide the error of the current iterate into two parts, called smooth and nonsmooth or sometimes low and high frequency. The latter part is taken care of by a so-called smoother **S** and the other part by the coarse grid correction, which solves the suitably transformed problem in a space with fewer unknowns using the same approach again, thus leading to a recursive method on multiple grids. The point here is to choose the coarse space such that the smooth error can be represented well in that space and thus the dimension of the problem has been significantly decreased.

There are basically three concepts to choose the coarse space. The first is the original one, where the computational grid is coarsened in some way. Then there are algebraic multigrid methods, which just use algebraic operations to reduce the number of unknowns. These have been generalized to the purely algebraic notion of multilevel methods. Finally, in particular for DG methods, there are multi-p methods (also called p-multigrid), that use lower order discretizations to reduce the space dimension.

Here, we assume that the nonlinear equation system arises from the method of lines, implying that the coarsening happens on the spatial grid. For space-time discretizations, there is the option to use a space-time multigrid, which includes coarsening in time.

We now first explain the multigrid method for linear problems, before looking at nonlinear problems and then multi-p methods. After that, we will demonstrate the benefit of designing a multigrid scheme directly for unsteady flows on a model equation and RK smoothers.

To describe the method, assume that a hierarchy of spaces is given, denoted by their level l, where a smaller index corresponds to a smaller space. Based on this, a restriction operator $\mathbf{R}_{l-1,l}$ to go from level l to the next coarse level $l-1$ and a prolongation operator $\mathbf{P}_{l,l-1}$ for the return operation are defined.

5.6.1   Multigrid for linear problems

As a model problem, we consider the linear advection equation

$$u_t + au_x = 0 \tag{5.24}$$

with $a > 0$ on the interval $x \in [0, 1]$ with periodic boundary conditions. The eigenfunctions of the spatial operator are $\phi(x) = e^{2\pi kix}$ for $k \in \mathbb{Z}$, which are again periodic and correspond to different frequencies.

A finite volume method discretization with equidistant mesh width Δx leads to the evolution equation for the cell average u_i in one cell i:

$$u_{i_t} + \frac{a}{\Delta x}(u_i - u_{i-1}) = 0.$$

Using the vector $\mathbf{u} = (u_1, ..., u_m)^T$ and

$$\mathbf{B} = \begin{pmatrix} 1 & & & & -1 \\ -1 & 1 & & & \\ & -1 & 1 & & \\ & & \ddots & \ddots & \\ & & & -1 & 1 \end{pmatrix},$$

we obtain the system of ODEs

$$\mathbf{u}_t + \frac{a}{\Delta x}\mathbf{B}\mathbf{u}(t) = \mathbf{0}. \tag{5.25}$$

Here, we discretize this using implicit Euler with time step size Δt, obtaining the linear system

$$\mathbf{u}^{n+1} - \mathbf{u}^n + \frac{a\Delta t}{\Delta x}\mathbf{B}\mathbf{u}^{n+1} = 0$$

$$\Leftrightarrow \mathbf{u}^n - \mathbf{A}\mathbf{u}^{n+1} = 0, \tag{5.26}$$

where

$$\mathbf{A} = \left(\mathbf{I} + \frac{\nu}{\Delta x}\mathbf{B}\right) \tag{5.27}$$

with $\nu = a\Delta t$. Here, $CFL := a\Delta t/\Delta x$ corresponds to the CFL number in the implicit Euler method. If we consider nonperiodic boundary conditions, the entry in the upper right corner of $\mathbf{B}$ becomes zero. Furthermore, additional terms appear on the right-hand side, but this does not affect multigrid convergence.

Now, an important property is that the eigenvectors of the matrix $\mathbf{B}$ are discrete forms of the eigenfunctions $e^{2\pi ikx}$ of $\partial/\partial x$, obtained by sampling those in $x_j = j\Delta x$ for the m frequencies $k = -n, ..., n$. Here, it is useful to use the phase angles $\Theta = 2\pi k/n$, obtaining the possible eigenvectors $\{e^{i\Theta j}\}_{j=1,...,m}$ for $\Theta \in [-\pi, \pi]$. Finally, the eigenvalues are given by $1 - e^{-i\Theta}$. If nonperiodic boundary conditions are used, the matrix becomes lower triangular and all eigenvalues are equal to $-1 - \frac{\nu}{\Delta x}$. However, since the matrix is nonnormal, the eigenvalues and eigenvectors cannot be used to do a stability analysis. Correspondingly, the eigenvalues of $\mathbf{A}$ are given by

$$\lambda(\Theta) = 1 + \frac{\nu}{\Delta x}(1 - e^{-i\Theta}). \tag{5.28}$$

Now the point is that if we define a coarse grid by dropping every other point, then only the slowly varying low frequency error components ($|\Theta| \in [0, \pi/2]$) can be represented there. On the other hand, the high frequency components ($|\Theta| \in [\pi/2, \pi]$) cannot be represented and need to be eliminated by a smoother.

If a different linear PDE is considered, the design of a multigrid method for that equation repeats the steps from above: First determine the eigenfunctions of the continuous operator and their discrete counterparts for the specific discretization given. Then define the coarse space and determine the eigenfunctions/eigenvectors present in the coarse space. With that information, find a smoother that takes care of the other components of the error.

This means that the notions of high and low frequency errors depend on the equation to be solved and the discretization used. Correspondingly, the components of the method (smoother, restriction, and prolongation) have to be chosen in an appropriate way and the same smoother may not work for a different equation. In the case that we do not consider a linear, but a nonlinear PDE and the linear equation system comes from a linearization, e.g., from Rosenbrock time integration or Newton's method, the design of multigrid methods in this way becomes even more difficult.

For conservation and balance laws, it is important to have restriction and prolongation operators that are conservative. To this end, a coarse grid is obtained by agglomerating a number of neighboring cells, giving rise to the term "agglomeration multigrid." The restricted value in a coarse cell is then given by summing up the fine grid values, weighted by the volumes of the respective cells and dividing by the total volume to obtain a conservative restriction. For an equidistant grid in one dimension, the corresponding restriction operator would be given by

$$\mathbf{R}_{l-1,l} = \frac{1}{2} \begin{pmatrix} 1 & 1 & & & \\ & & 1 & 1 & \\ & & & & \ddots & \\ & & & & & 1 & 1 \end{pmatrix}.$$

One possible prolongation is the injection, where the value in the coarse cell is taken as value on all the corresponding fine cells, which would result in

$$\mathbf{P}_{l,l-1} = 2\mathbf{R}_{l-1,l}^T. \tag{5.29}$$

This can be refined by interpolating means of reconstructed functions. Jameson suggests a bilinear interpolation, which in addition to the coarse grid point corresponding to a fine grid point incorporates those neighboring coarse grid cells which would incorporate a fine grid neighbor of the fine grid point considered, using suitable weights.

Furthermore, the problem on the coarse grids has to be defined. A common way for linear problems is to obtain the matrix by the so-called Galerkin condition

$$\mathbf{A}_{l-1} = \mathbf{R}_{l-1,l}\mathbf{A}_l\mathbf{P}_{l,l-1}.$$

Alternatively, the problem can be discretized directly on the coarse level, which will be the way to go for nonlinear problems.

To obtain the right-hand side on the coarse level, we now use that for linear problems if $\mathbf{A}\mathbf{x} - \mathbf{b} = \mathbf{r}$, then

$$\mathbf{A}\mathbf{e} = \mathbf{r}. \tag{5.30}$$

This is called the defect equation and means that on the coarse level, we solve for the error and then correct the fine level solution by the prolongated error on the coarse level, which completes the coarse grid correction.

Summing up, we obtain the multigrid method to solve the system $\mathbf{A}\mathbf{x} = \mathbf{b}$:

Function MG$(\mathbf{x}_l, \mathbf{b}_l, l)$

- if $(l = 0)$, $\mathbf{x}_l = \mathbf{A}_l^{-1}\mathbf{b}_l$ (Exact solve on coarse grid)

- else

    - $\mathbf{x}_l = \mathbf{S}_l^{\nu_1}(\mathbf{x}_l, \mathbf{b}_l)$ (Presmoothing)
    - $\mathbf{r}_{l-1} = \mathbf{R}_{l-1,l}(\mathbf{b}_l - \mathbf{A}_l\mathbf{x}_l)$ (Restriction of Residual)
    - $\mathbf{v}_{l-1} = 0$
    - For $(j = 0; \; j < \gamma; \; j++)$ MG$(\mathbf{v}_{l-1}, \mathbf{r}_{l-1}, l - 1)$ (Computation of the coarse grid correction)
    - $\mathbf{x}_l = \mathbf{x}_l + \mathbf{P}_{l,l-1}\mathbf{v}_{l-1}$ (Correction via Prolongation)
    - $\mathbf{x}_l = \mathbf{S}_l^{\nu_2}(\mathbf{x}_l, \mathbf{b}_l)$ (Postsmoothing)

- end if

If $\gamma = 1$, we obtain a so-called V-cycle, see Figure 5.3, where each grid is visited only once. Choosing $\gamma = 2$ results in a W-cycle, see Figure 5.4. Larger choices of γ will not be used, since there is rarely a benefit for the additional cost.

5.6.2   Full Approximation Schemes

If instead of a linear problem, a nonlinear equation

$$\mathbf{f}(\underline{\mathbf{u}}) = \mathbf{s}$$

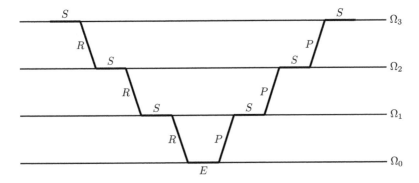

FIGURE 5.3: V-multigrid cycle. (Credit: L. M. Versbach.)

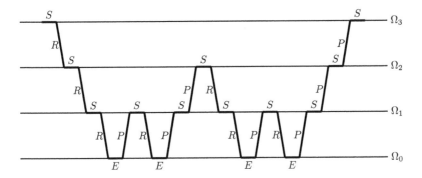

FIGURE 5.4: W-multigrid cycle. (Credit: L. M. Versbach.)

is considered, the multigrid method has to be modified and is called Full Approximation Scheme (FAS). This type of method is widespread in industry, in particular for steady flows, where textbook multigrid efficiency has been demonstrated for the steady Euler equations [55] and at least mesh width independent convergence rates for the Navier-Stokes equations. The point where we have to modify the scheme is that for a nonlinear operator, we have in general

$$\mathbf{f}(\underline{\mathbf{u}}) - \mathbf{f}(\underline{\mathbf{u}}^*) \neq \mathbf{f}(\mathbf{e})$$

and therefore the defect equation (5.30) is not valid and we cannot solve for the coarse grid error directly. Therefore, we instead consider the equation

$$\mathbf{f}(\underline{\mathbf{u}}^*) = \mathbf{f}(\underline{\mathbf{u}}) - \mathbf{r}(\underline{\mathbf{u}}), \tag{5.31}$$

which is equivalent to the defect equation in the linear case. On the coarse grid we thus solve

$$\mathbf{f}_{l-1}(\underline{\mathbf{u}}_{l-1}) = \mathbf{s}_{l-1} := \mathbf{f}_{l-1}(\mathbf{R}_{l-1,l}\underline{\mathbf{u}}_l) + \mathbf{R}_{l-1,l}(\mathbf{s}_l - \mathbf{f}_l(\underline{\mathbf{u}}_l)). \tag{5.32}$$

Due to the nonlinearity, the right-hand side might be such that it is not in the image of $\mathbf{f}$. In this case, the residual needs to be damped by some factor α. However, this does not seem to be necessary in our case. There is one more change, because even on the coarsest level, it is impossible to solve the nonlinear equation exactly. Therefore, only a smoothing step is performed there. The prolongation $\mathbf{Q}_{l,l+1}$ and restriction $\mathbf{R}_{l+1,l}$ are chosen as described in the last section and the coarse grid problem is defined in a natural way as if the equation would be discretized on that grid. We thus obtain the following method:

Function FAS-MG($\underline{\mathbf{u}}_l, \mathbf{s}_l, l$)

- $\underline{\mathbf{u}}_l = \mathbf{S}_l^{\nu_1}(\underline{\mathbf{u}}_l, \mathbf{s}_l)$ (Presmoothing)

- if ($l > 0$)

    - $\mathbf{r}_l = \mathbf{s}_l - \mathbf{f}_l(\underline{\mathbf{u}}_l)$
    - $\tilde{\mathbf{u}}_{l-1} = \mathbf{R}_{l-1,l}\underline{\mathbf{u}}_l$ (Restriction of solution)
    - $\mathbf{s}_{l-1} = \mathbf{f}_{l-1}(\tilde{\mathbf{u}}_{l-1}) + \mathbf{R}_{l-1,l}\mathbf{r}_l$ (Restriction of residual)
    - For ($j = 0$; $j < \gamma$; $j++$) FAS-MG($\underline{\mathbf{u}}_{l-1}, \mathbf{s}_{l-1}, l-1$) (Computation of the coarse grid correction)
    - $\underline{\mathbf{u}}_l = \underline{\mathbf{u}}_l + \mathbf{P}_{l,l-1}(\underline{\mathbf{u}}_{l-1} - \tilde{\mathbf{u}}_{l-1})$ (Correction via Prolongation)
    - $\underline{\mathbf{u}}_l = \mathbf{S}_l^{\nu_2}(\underline{\mathbf{u}}_l, \mathbf{s}_l)$ (Postsmoothing)

- end if

5.6.3   Smoothers

The choice of smoother is crucial for the success of the method. Different methods have been successfully used, namely explicit RK schemes, explicit additive RK methods, point implicit smoothers, line implicit smoothers, and the SGS method.

5.6.3.1   Pseudo time iterations and dual time stepping

Regarding time integration schemes as smoothers, their relation to iterative solvers first has to be understood. Consider the algebraic equation

$$\mathbf{f}_l(\underline{\mathbf{u}}_l) = \mathbf{s}_l, \tag{5.33}$$

with solution $\underline{\mathbf{u}}_l^*$, to be solved on level l of the multigrid method. Now we make the variable $\underline{\mathbf{u}}_l$ to be dependent on a pseudo time t^* and for a given initial guess $\underline{\mathbf{u}}^0$, we look at the initial value problem

$$\frac{\partial \underline{\mathbf{u}}(t^*)}{\partial t^*} + \mathbf{f}_l(\underline{\mathbf{u}}_l(t^*)) - \mathbf{s}_l = \mathbf{0}, \quad \underline{\mathbf{u}}(t_0^*) = \underline{\mathbf{u}}^0. \tag{5.34}$$

The idea is now that as long as this IVP has a steady state with

$$\mathbf{u}_l^* = \lim_{t^* \to \infty} \mathbf{u}_l(t^*),$$

then we can use any convergent numerical time integration method as an iterative method to compute the solution of (5.33). The time step index then denotes an iteration index.

These methods originated from the computation of steady states, with the physically motivated idea of marching in time toward the steady state, similar to a homotopy method. However, Jameson realized that this can be used to compute solutions of algebraic equations arising within implicit time integration as well, and that the FAS method for the computation of steady states can be used to do this with a minimal amount of code changes. For the case of the implicit Euler method as in (5.1), we obtain that the flow of the IVP has to be changed to $\mathbf{u}_l(t^*) - \Delta t \underline{\mathbf{f}}_l(\mathbf{u}(t^*))$. Note that using the opposite sign will lead to an IVP exhibiting blowup. He called this the dual time stepping approach [142]. In contrast to steady flow computations, the physical time plays an inherent role in unsteady computations. Thus, Jameson came up with the notion of dual or pseudo time t^* only for unsteady flows.

Thus, in the CFD community, the notion of dual time stepping is inherently connected to the nonlinear multigrid method of Jameson, and is usually meant as a nonlinear multigrid method to solve algebraic equations within implicit time integration for unsteady flows. However, the term becomes confusing when the smoother employed is not a time integration method like ARK, but another iterative method where no pseudo time is employed. Therefore, it is more clear when one speaks of FAS for steady or unsteady equations with a specific smoother, which can be a pseudo time iteration.

In the case of a linear equation with flux $\mathbf{f}(\underline{\mathbf{u}}) = \mathbf{A}\mathbf{u}$, pseudo time iterations are stationary linear methods. The iteration matrix is then the stability function evaluated in $-\Delta t^* \mathbf{A}$.

Calling the initial value $\mathbf{u}_l^{(0)}$ and using a low storage explicit RK method (4.18) for equation (5.34), we obtain the scheme

$$\mathbf{u}_l^{(1)} = \mathbf{u}_l^{(0)} - \alpha_1 \Delta t_l^* [\mathbf{f}_l(\mathbf{u}^{(0)}) + \mathbf{s}_l]$$

$$\dots$$

$$\mathbf{u}_l^{(q+1)} = \mathbf{u}_l^{(0)} - \alpha_{q+1} \Delta t_l^* [\mathbf{f}_l(\mathbf{u}^{(q)}) + \mathbf{s}_l]$$

$$\mathbf{u}_l^{n+1} = \mathbf{u}_l^{(q+1)},$$

where Δt^* is the time step in pseudo time t^*.

In the seminal paper [146], Jameson, Schmidt, and Turkel suggested the use of additive RK methods (4.27), where different coefficients are used for the convective and the diffusive parts. This is done to reduce the number of evaluations of the expensive diffusive terms, but also to increase the degrees of freedom in choosing a good smoother. Given a splitting in the convective and diffusive part

$$\mathbf{f}(\underline{u}) = \mathbf{f}^c(\underline{u}) + \mathbf{f}^v(\underline{u}),$$

the following equivalent form is written down in [146]:

$$\underline{u}_l^{(0)} = \underline{u}_l$$
$$\underline{u}_l^{(j)} = \underline{u}_l - \alpha_j \Delta t_l^* (\mathbf{f}^{c,(j-1)} + \mathbf{f}^{v,(j-1)}), \quad j = 1, ..., s$$
$$\underline{u}_l^{n+1} = \underline{u}_l^{(s)},$$

where

$$\mathbf{f}^{c,(j)} = \mathbf{f}^c(\underline{u}_l^{(j)}), \quad j = 0, ..., s-1$$
$$\mathbf{f}^{v,(0)} = \mathbf{f}^v(\underline{u}_l^{(0)}),$$
$$\mathbf{f}^{v,(j)} = \beta_j \mathbf{f}^v(\underline{u}_l^{(j)}) + (1 - \beta_j)\mathbf{f}^{v,(j-1)}, \quad j = 1, ..., s-1.$$

This is also called a JST type smoother. The methods coefficients are then chosen to provide good damping of high frequency modes. For the one-dimensional linear advection equation

$$u_t + au_x = 0, \tag{5.35}$$

with $a > 0$, discretized using a first-order upwind scheme with fixed mesh width Δx and explicit RK smoothing, this was done in [287]. However, the methods obtained in this way are not convincing when applied to the Euler or Navier-Stokes equations. Instead, Jameson used the one-dimensional linear advection with a fourth-order derivative to obtain a better smoother for Fourth-order artificial diffusion schemes [141]. The coefficients for the corresponding additive RK smoothers can be found in Table 5.1 and [143]. Furthermore, the relation to additive RK methods as written down in (4.27) is given in Tables 5.2 and 5.3, where the corresponding Butcher arrays are described.

i	1	2	3	4	5
α_i	1/3	4/15	5/9	1	-
β_i	1	1/2	0	0	-
α_i	1/4	1/6	3/8	1/2	1
β_i	1	0	0.56	0	0.44

TABLE 5.1: Coefficients of additive Runge-Kutta smoothers, 4-stage and 5-stage method.

While this class of schemes was very influential, they are too slow in practice. The main reason is that they are explicit at the end, with correspondingly little global coupling. Consequently, faster methods are more implicit in nature. A very fruitful development resides under the name preconditioned pseudo time iterations or additive W (AW) smoothers [237, 269, 238, 166,

$$
\begin{array}{c|cccc}
0 & 0 & 0 & 0 \\
\alpha_1 & 0 & 0 & 0 \\
0 & \ddots & \ddots & 0 \\
0 & 0 & \alpha_{s-1} & 0 \\
\hline
0 & \cdots & 0 & \alpha_{s-1}
\end{array}
$$

TABLE 5.2: Conversion of coefficients of JST type smoother into Butcher array: Convective terms.

$$
\begin{array}{c|ccccc}
0 & & \cdots & & & 0 \\
\alpha_1 & 0 & & & & 0 \\
\alpha_2(1-\beta_1) & \alpha_2\beta_2 & \ddots & & & 0 \\
0 & \ddots & \ddots & & 0 & 0 \\
0 & \cdots & & \alpha_s(1-\beta_{s-1}) & \alpha_s\beta_s & 0 \\
\hline
0 & \cdots & & 0 & \alpha_s(1-\beta_{s-1}) & \alpha_s\beta_s
\end{array}
$$

TABLE 5.3: Conversion of coefficients of JST type smoother into Butcher array: Diffusive terms.

169, 268, 167, 168, 170, 144, 28, 29, 222, 30]. It comes down to, instead of the additive explicit Runge-Kutta method, using an additive W method, compare (4.28). Setting all additional coefficients γ_{ij} to zero gives us:

$$\mathbf{u}^{(0)} = \mathbf{u}^n$$
$$\mathbf{u}^{(i)} = \mathbf{u}^n - \alpha_i \Delta t^* \mathbf{W}^{-1}(\mathbf{f}^{c,(i-1)} + \mathbf{f}^{v,(i-1)}), \quad i = 1, ..., s$$
$$\mathbf{u}^{n+1} = \mathbf{u}^{(s)},$$

whereas before

$$\mathbf{f}^{c,(i)} = \mathbf{f}^c(\mathbf{u}^{(i)}), \quad i = 0, ..., s-1$$
$$\mathbf{f}^{v,(0)} = \mathbf{f}^v(\mathbf{u}^{(0)}),$$
$$\mathbf{f}^{v,(i)} = \beta_{i+1}\mathbf{f}^v(\mathbf{u}^{(i)}) + (1 - \beta_{i+1})\mathbf{f}^{v,(i-1)}, \quad i = 1, ..., s-1.$$

Here, $\mathbf{W} \approx \mathbf{I} + \Delta t^* \eta df(\mathbf{u}^{(0)})/d\mathbf{u}$. A choice that works very well is to use one or a few steps of SGS (see 5.17). A method that is easy to implement is as follows.

Basis for the Jacobian approximation is a different first-order linearized discretization of the Euler equations. It is based on a splitting $\mathbf{A} = \mathbf{A}^+ + \mathbf{A}^-$ of the flux Jacobian. This is evaluated in the average of the values on both sides of the interface. The split Jacobians correspond to positive and negative eigenvalues:

$$\mathbf{A}^+ = \frac{1}{2}(\mathbf{A} + |\mathbf{A}|), \quad \mathbf{A}^- = \frac{1}{2}(\mathbf{A} - |\mathbf{A}|).$$

with $|\mathbf{A}| := \mathbf{R}|\Lambda|\mathbf{R}^{-1}$. Alternatively, these can be written in terms of the matrix of right eigenvectors $\mathbf{R}$ as

$$\mathbf{A}^+ = \mathbf{R}|\Lambda^+|\mathbf{R}^{-1}, \quad \mathbf{A}^- = \mathbf{R}|\Lambda^-|\mathbf{R}^{-1},$$

where $\Lambda^\pm$ are diagonal matrices containing the positive and negative eigenvalues, respectively. Then, a cutoff function to bound the eigenvalues away from zero [270, 140] is used. One possibility is a parabolic function

$$|\lambda| = \frac{1}{2}\left(cd + \frac{|\lambda|^2}{cd}\right), \quad |\lambda| \leq cd \tag{5.36}$$

that uses a new parameter $d \in [0,1]$ representing a fraction of the speed of sound, c.

With this, an upwind discretization is given in cell i by

$$\mathbf{u}_{i_t} = \frac{3}{2\Delta t^*}\mathbf{I} + \frac{1}{\Omega_i}\sum_{e_{ij}\in N(i)} |e_{ij}|(\mathbf{A}^+\mathbf{u}_i + \mathbf{A}^-\mathbf{u}_j). \tag{5.37}$$

The Jacobian of this discretization is then the basis for the SGS preconditioner with blocks

$$\mathbf{L}_{ij} = -\frac{\eta\Delta t_i^*}{\Omega_i}(\Delta y\mathbf{A}_{i-1,j}^+ + \Delta x\mathbf{B}_{i,j-1}^+), \tag{5.38}$$

$$\mathbf{U}_{ij} = \frac{\eta\Delta t_i^*}{\Omega_i}(\Delta y\mathbf{A}_{i-1,j}^- + \Delta x\mathbf{B}_{i,j-1}^-), \tag{5.39}$$

$$\mathbf{D}_{ii} = \mathbf{I} + \frac{3\eta\Delta t^*}{2\Delta t}\mathbf{I} + \frac{\eta\Delta t_i^*}{\Omega_i}[\Delta y(\mathbf{A}_{ii}^+ - \mathbf{A}_{ii}^-) + \Delta x(\mathbf{B}_{ii}^+ - \mathbf{B}_{ii}^-)]. \tag{5.40}$$

Applying this preconditioner requires solving small 4×4 systems coming from the diagonal, which can be done efficiently using Gaussian elimination. A fast implementation is obtained by transforming first to a certain set of symmetrizing variables (see [269]). Since the underlying method is implicit, Δt^* can be chosen arbitrarily large. With regards to the other two new parameters, η and c, a robust choice is $\eta = 0.5$ and $d = 0.5$. Smaller values of d lead to faster convergence, but less robustness [30].

To make these schemes efficient, a number of additional techniques can be used. A crucial one is local time stepping. Since the pseudo time has no physical meaning, it is possible to use a different value for Δt^* in every cell. This is typically implemented by defining a pseudo cfl number c^* and then computing the local pseudo time step based on the local Δx. This can be used both on unstructured meshes, but also on coarse grids, so that really large time steps are employed.

Another technique is preconditioning of the time derivative. This consists of multiplying the pseudo time derivative with a preconditioner, i.e., a nonsingular matrix. Since the assumption is that the pseudo time derivative vanishes in the limit, this does not change the solution of the original algebraic equation, similar to the preconditioners for linear systems discussed later. An example is the above local time stepping, which corresponds to preconditioning with a diagonal matrix.

5.6.3.2 Alternatives

A powerful alternative to RK smoothers is the SGS method (5.17) [55]. This is typically used in conjunction with flux vector splitting discretizations, since this results in significantly less computational work.

Point implicit smoothers correspond to using an SDIRK scheme with only one Newton step, where the Jacobian is approximated by a block diagonal matrix [166, 169]. Line implicit smoothers are used to cope with grid-induced stiffness by defining lines in anisotropic regions normal to the direction of stretching. These lines then define a set of cells, for which an SDIRK method is applied, again with one Newton step, but taking into account the coupling between the cells along the line only. This leads to a tridiagonal block matrix [190, 167]. Another idea is to solve the linear system arising within Newton using preconditioned GMRES [170], or to use GMRES as a smoother [29, 176], which are discussed in sections 5.7 and 5.8. This gives promising results. Unfortunately, these schemes are very hard to analyze.

5.6.4 Residual averaging and smoothed aggregation

Residual averaging is used to increase the level of implicitness in the method and thus to move unphysical parts of the solution even faster out of the domain [145]. To this end, a small parameter ϵ is chosen and then a Laplace-filter is applied directionwise:

$$- \epsilon \bar{r}_{j-1} + (1 + 2\epsilon)\bar{r}_j - \epsilon \bar{r}_{j+1} = r_j. \tag{5.41}$$

This equation is solved for the new residuals $\bar{r}$ using a Jacobi iteration. For a suitably large parameter ϵ, this stabilizes the scheme.

Another technique is smoothed aggregation, which was applied to flow problems, for example in [243]. Here, an aggregation multigrid using the prolongation (5.29) is augmented by a damped Jacobi smoothing step after the prolongation, giving

$$\tilde{\mathbf{P}}_{l,l-1} = (\mathbf{I} - \omega \mathbf{D}_l^{-1} \mathbf{A}_l)\mathbf{P}_{l,l-1}.$$

For symmetric problems, the restriction would then be chosen as the transpose of this. For unsymmetric problems, it is better to leave the restriction as is, leading to the method nonsmoothed restriction (NSR).

5.6.5   Multi-p methods

Another alternative in the context of DG and other higher order methods are multi-p schemes, where instead of coarsening the grid, the coarse scale problem is defined locally on each element using a lower order method [90]. Therefore, these two approaches are also often called p- and h-multigrid. Mesh width independent convergence rates have been demonstrated numerically for steady state problems. To define the method, let a high order space $\mathbf{\Pi}_l$ and a nested low order space $\mathbf{\Pi}_{l-1} \subset \mathbf{\Pi}_l$ be given. Typically, the low order space is defined by decreasing the polynomial order by 1, although in the original method, a more aggressive coarsening is used, dividing the order by 2. The most significant difference is in the size of the coarse subspaces. Then, the FAS multigrid method is applied as before, where we have to define new prolongations and restrictions. For the restriction of the solution (represented by the coefficients of a polynomial ansatz) and the residual, different operators $\mathbf{R}_{l-1,l}$ and $\tilde{\mathbf{R}}_{l-1,l}$ are used, since these objects live in different spaces.

To interpolate the solution from the lower to the higher order space, we use that there is a unique representation of the basis functions ϕ_i^{l-1} of the low order space in the basis of the high order space:

$$\phi_i^{l-1} = \sum_j \alpha_{ij} \phi_j^l.$$

Then, the prolongation is given by $\mathbf{P}_{l,l-1} = (\alpha_{ij})_{ij}$. For a hierarchical basis, this representation is trivial, in that the prolongation is just an identity matrix with added zero columns.

For the restriction of the residual, $\tilde{\mathbf{R}}_{l-1,l} = \mathbf{P}_{l,l-1}^T$ is typically used, which corresponds to a Galerkin type definition of the coarse space. The restriction of the states can be defined by an L_2-projection, which has the advantage of being conservative. Thus we require

$$(\mathbf{w}^{l-1}, \mathbf{v}^{l-1}) = (\mathbf{w}^{l-1}, \mathbf{v}^l), \ \forall \mathbf{w}^{l-1} \in \mathbf{\Pi}_{l-1}.$$

This leads to

$$\mathbf{R}_{l,l-1} = \mathbf{M}_{l,l-1}^{-1} \mathbf{N}_{l,l-1},$$

where $(\mathbf{M}_{l,l-1})_{ij} = (\phi_i^{l-1}, \phi_j^{l-1})$ is a quadratic matrix and $(\mathbf{N}_{l,l-1})_{ij} = (\phi_i^{l-1}, \phi_j^l)$ is rectangular. For a DG method with an orthogonal basis, as for example the modal-nodal method described in section 3.7.1, we have $(\phi_i^p, \phi_j^p) = \delta_{ij}$ and thus $\mathbf{M}_{l,l-1} = \mathbf{I}$. If a nodal basis is used, $\mathbf{M}_{l,l-1}$ is dense.

Regarding the smoother, because the coarse grid has been defined differently from before, their performance has to be assessed anew. Alternatives that have been used are element-Jacobi [298], line-Jacobi [90], ILU [224], RK methods [14, 151], and combinations thereof [182]. In the context of the linear advection equation, optimal RK smoothers were analyzed in [13]. There, coefficients and corresponding optimal time step sizes are computed for the case of steady flows.

5.6.6   State of the art

In the context of compressible flows, the most widely used and fastest FAS type scheme is the one developed by Jameson over the course of 30 years [143]. For the Euler equations, the latest version is due to Jameson and Caughey [55] and solves the steady Euler equations with a finite volume discretization with the JST flux function around an airfoil in three to five steps. Thus, two-dimensional flows around airfoils can be solved on a PC in a matter of seconds. The solution of the RANS equations is more difficult, but with the recently developed AW smoothers only between 10 and 200 steps for engineering accuracies are needed and only a weak dependency on mesh resolution is reported, even in 3D [269, 170, 30]. Thus, we are close to textbook multigrid efficiency for the RANS equations.

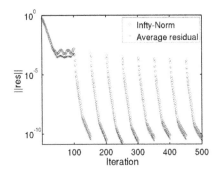
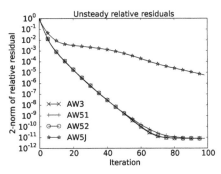

FIGURE 5.5: Comparison of convergence of FAS multigrid in UFLO103 for two-dimensional flow around a pitching and plunging NACA 64A010 airfoil at Mach number 0.796. Left: For Euler equations. Right: For the RANS equations. (Credit: From [30], CC-by-SA 4.0 (https://creativecommons.org/licenses/by-sa/4.0/). No changes were made.)

In Figure 5.5, we demonstrate the convergence behavior of the FAS multigrid scheme implemented in the UFLO103 code of Jameson. The test case is two-dimensional flow around a pitching and plunging NACA 64A010 airfoil at a Mach number 0.796. For the pitching, we use a frequency of 0.202 and an amplitude of 1.01°. Thirty-six time-steps per cycle (pstep) are chosen. The Reynolds number is 10^6 and the Prandtl number is 0.75 and the turbulence model is the 0-equation Baldwin-Lomax model. Regarding the grid, a C-mesh with 512×64 cells and maximum aspect ratio of 6.31e6 is employed. The multigrid method uses a W-cycle with five grid levels. On the left, we show an Euler computation over several time steps. The first 100 iterations are the startup phase, where the steady state around the airfoil is computed to get initial conditions for the unsteady computation. From then, 50 FAS iterations are done per time step. The reduction in residual is about 8 orders of magnitude. On the right, we see the convergence history for one time step of a URANS computation. Using AW5, we can reduce the residual by 11 orders of magnitude in 70 iterations.

Regarding LES computations with high order methods, textbook multigrid efficiency has not yet been demonstrated. Currently, the state of the art uses the later discussed preconditioned Newton-Krylov solvers.

5.6.7   Analysis and construction

Regarding convergence theory, note that for linear problems, we have a stationary linear method (see section 5.5.1). The iteration matrix of the two-grid scheme with one presmoothing step and without postsmoothing is

$$\mathbf{M}_{2G_l} = (\mathbf{I} - \mathbf{P}_{l,l-1}\mathbf{A}_{l-1}^{-1}\mathbf{R}_{l-1,l}\mathbf{A}_l)\mathbf{M}_{S_l} \tag{5.42}$$

and we have

$$\begin{aligned}
\mathbf{N}_{2G_l}^{-1} &= (\mathbf{I} - \mathbf{P}_{l,l-1}\mathbf{A}_{l-1}^{-1}\mathbf{R}_{l-1,l}\mathbf{A}_l)\mathbf{N}_{S_l}^{-1} + \mathbf{P}_{l,l-1}\mathbf{A}_{l-1}^{-1}\mathbf{R}_{l-1,l} \\
&= \mathbf{N}_{S_l}^{-1} + \mathbf{P}_{l,l-1}\mathbf{A}_{l-1}^{-1}\mathbf{R}_{l-1,l}(\mathbf{I} - \mathbf{A}_l\mathbf{N}_{S_l}^{-1}). \tag{5.43}
\end{aligned}$$

Here, $\mathbf{M}_S$ and $\mathbf{N}_{S_l}^{-1}$ and the matrices defining the smoother as a stationary linear method. In a multigrid method, $\mathbf{A}_{l-1}^{-1}$ is replaced by a call to the two-grid method with initial guess $\mathbf{0}$. Thus, for a V-cycle this would be $\mathbf{N}_{2G_{l-1}}^{-1}$.

Theorem 6 says that the method converges if $\rho(\mathbf{M}) < 1$. The common approach to prove this and in particular that $\rho(\mathbf{M})$ is independent of the mesh width, is the analysis as established by Hackbusch [120], which distinguishes between the smoothing and the approximation property. The first one describes if the smoothing operator is able to deal with the high frequency errors and the approximation property describes if the coarse grid solution is able to approximate the remaining components of the error.

The convergence theory for elliptic problems is well established and significant progress on the design of multigrid methods has been made for the incompressible Navier-Stokes equations [282]. By contrast, there is very little theory for the compressible Navier-Stokes equations. This leads to a situation, where experience and intuition play as large a role as analysis.

5.6.7.1   Smoothing factors

A straightforward notion for the construction of smoothers arises from the idea that the free parameters of the smoother should be chosen, such that error reduction for the high frequency part HF of the spectrum of the equation is optimized. The smoothing factor is then the by modulus largest of the high frequency parts of the eigenvalues λ_i^S of the iteration matrix $\mathbf{M}_S$ of the smoother:

$$\max_{i \in HF} |\lambda_i^S|. \tag{5.44}$$

It is not indicative of two-grid convergence rate, but gives nevertheless a reasonable recipe to construct smoothers, which works well in practice.

5.6.7.2 Example

We now illustrate this on the example of s-stage explicit RK smoothers for a model discretization, namely the one-dimensional linear advection equation with an upwind discretization (see section 5.6.1 and equation (5.24)). Van Leer et al. discussed finding a smoother, with optimal, meaning minimal, smoothing factor [287]. Since the iteration matrix of such smoothers is represented by the degree s polynomial P_s, this comes down to the optimization problem

$$\min_{\Delta t, \alpha} \max_{|\Theta| \in [\pi/2, \pi]} |P_s(z(\Theta))|, \qquad (5.45)$$

where α is the set of free coefficients of the polynomial and $z(\Theta)$ are the eigenvalues of the matrix $\mathbf{A}$ from (5.27). For this case, the optimization problem can be solved directly, but for more complex equations or smoothers, this is not possible. This approach was later also followed in [13, 26]. We follow [26].

Due to the linearity, it is sufficient to look at $P_s(\Delta t^* \lambda(\Theta))$ with

$$\Delta t^* \lambda(\Theta) = -\Delta t^* - \frac{\nu \Delta t^*}{\Delta x}(1 - e^{-i\Theta}).$$

Possible parameters of the smoother are the pseudo time step size Δt^* and the coefficients of the RK method. Now, $\nu = a\Delta t$ is fixed during the multigrid iteration, but Δx is not. Furthermore, the pseudo time step is restricted by a CFL condition based on ν. Thus, instead of optimizing for Δt^*, we define the pseudo time step on each grid level as

$$\Delta t_l^* = c \Delta x_l / \nu$$

and optimize for $c := \nu \Delta t_l^* / \Delta x_l$, implying that we also use local time stepping within the pseudo time iteration. Now we have

$$z(\Theta, c; \nu, \Delta x_l) := \Delta t^* \lambda(\Theta) = -c\Delta x_l / \nu - c + ce^{-i\Theta}, \qquad (5.46)$$

where we see that z does not depend on ν and Δx_l separately, but only on $\Delta x_l / \nu = 1/CFL$. Thus, with $e^{-i\Theta} = \cos(\Theta) - i\sin(\Theta)$ we obtain

$$z(\Theta, c; CFL) = -c/CFL - c + c\cos(\Theta) - ic\sin(\Theta).$$

In the end, given CFL, we have to solve the optimization problem (5.45). Using symmetry of P_s and equivalently looking at the square of the modulus, we obtain

$$\min_{c, P_s} \max_{|\Theta| \in [\pi/2, \pi]} |P_s(z(\Theta, c; CFL))|^2. \qquad (5.47)$$

Due to the dependence of the optimal coefficients on CFL, there is no unique optimal smoother for all problems.

For example, for the 2-stage scheme we have:

$$|P_2(z)|^2 = |1 + z + \alpha_1 z^2|^2$$
$$= (1 + \text{Re}z + \alpha_1 \text{Re}z^2 - \alpha_1 \text{Im}z^2)^2 + (2\alpha_1 \text{Re}z\text{Im}z + \text{Im}z)^2.$$

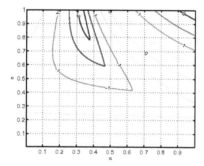

FIGURE 5.6: Contourplot of $\log_{10} \max_{|\Theta| \in [\pi/2, \pi]} |P_2(z(\Theta, c; 24))|^2$.

#stages	α_1	α_2	α_3	c	Opt-value
2	0.33			0.98	0.10257
3	0.145	0.395		1.495	0.01772
4	0.0867	0.2133	0.433	1.84	0.004313

TABLE 5.4: Results of optimization of smoothing properties for CFL=24.

It turns out that for these functions, the final form of (5.47) is too difficult to solve exactly, in particular due to the min-max formulation. Therefore, we discretize the parameter space and compute an approximate solution. This requires a bounded region, which is already the case for Θ and the α_j, which are between 0 and 1. As for c, we know that any explicit RK scheme has a bounded stability region; therefore, we chose an upper bound for c, such that the optimal value for c is not on the boundary. As an example, the function

$$f(\alpha, c) := \log_{10} \max_{|\Theta| \in [\pi/2, \pi]} |P_2(z(\Theta, c; CFL))|^2$$

is shown in Figure 5.6 for $CFL = 24$. Note that the optimal c is not on the boundary, which means that the choice $c \in [0, 1]$ here is reasonable. Furthermore, we can see that for $c = 0$, we obtain a method with a smoothing factor of 1. This is correct, since this is a method with time step zero, meaning that the resulting smoother is the identity. For $\alpha = 0$, we obtain the explicit Euler method. This is also a possible smoother, but as can be seen it is less powerful. Furthermore, we can see the finite stability regions of the methods.

We now compute the optimal value for $CFL = 24$ for all schemes using a MATLAB/C++ code. The optimization gives results presented in Table 5.4. For the 2-stage scheme, we choose a grid of $200 \times 200 \times 200$ for the parameter space $\alpha_1 \times c \times t$. For the 3-stage scheme, we have one more parameter, which increases the dimension of the parameter space $\alpha_1 \times \alpha_2 \times c \times t$, which results in the grid $200 \times 200 \times (2 \cdot 200) \times 200$. As a restriction for c, we put $c \in [0, 2]$. The optimal value is decreased by a factor of about 500, suggesting that these schemes are significantly better than the 2-stage methods. Finally, for the 4-stage scheme, we chose a grid of the size $150 \times 150 \times 150 \times (2 \cdot 150) \times 100$ with $c \in [0, 2]$. A finer grid was not possible due to storage restrictions. The optimal value is decreased by a factor of 4 compared to the 3-stage schemes.

An important difference between the schemes obtained is the size of c, which is about 0.9 for the 2-stage case, 1.4 for the 3-stage case, and 1.9 for the 4-stage case, which suggests that one effect of allowing more freedom in the design of the smoother is that the stability region is enlarged. The stability regions of the optimal methods obtained are shown in Figure 5.7, emphasizing this point.

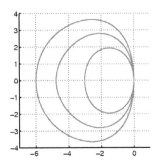

FIGURE 5.7: Stability regions of optimally smoothing methods $CFL = 24$. The larger the stability region, the higher the number of stages.

5.6.7.3 Optimizing the spectral radius

The optimization just considered aims at improving the smoother on its own, without taking into account the interaction with the coarse grid correction or the multigrid structure. This has the benefit that the optimization is fast, even for the 4-stage case, where we run into memory problems. An alternative is to optimize the spectral radius of the iteration matrix $\mathbf{M}$ of the complete multigrid scheme as a function of the smoother, which is a function of α and c, with c defined as above:

$$\min_{\alpha, c} \rho(\mathbf{M}(\alpha, c; \nu, \Delta x)). \tag{5.48}$$

The difference between the two approaches is that the latter one is theoretically able to provide truly optimal schemes, whereas the first one is not. However, the second approach is much more costly, which means that we are not able to compute the global optimum in a reasonable time.

5.6.7.4 Example

We return to the example from the previous section about finding good explicit RK smoothers. Since the example is linear, we have a stationary linear method, given by

$$S_{s,l}(\mathbf{b}, \mathbf{u}) = \mathbf{M}_{s,l}\mathbf{u} + \mathbf{N}_{s,l}^{-1}\mathbf{b},$$

where $\mathbf{M}_{s,l} = P_s(-\Delta t^* \mathbf{A}_l)$ is given by the stability polynomial, whereas the second matrix corresponds to a different polynomial. For the 2-, 3-, and 4-stage smoother, we have

$$\mathbf{N}_{2,l}^{-1} = \Delta t \mathbf{I}_l - \alpha_1 \Delta t^2 \mathbf{A}_l, \tag{5.49}$$

$$\mathbf{N}_{3,l}^{-1} = \Delta t \mathbf{I}_l - \alpha_1 \Delta t^2 \mathbf{A}_l + \alpha_1 \alpha_2 \Delta t^3 \mathbf{A}_l^2, \tag{5.50}$$

$$\mathbf{N}_{4,l}^{-1} = \Delta t \mathbf{I}_l - \alpha_1 \Delta t^2 \mathbf{A}_l + \alpha_1 \alpha_2 \Delta t^3 \mathbf{A}_l^2 - \alpha_1 \alpha_2 \alpha_3 \Delta t^4 \mathbf{A}_l^3. \tag{5.51}$$

We now consider a three-level scheme, whereas in FAS methods, only one smoothing step is done at the coarsest level. Then the iteration matrix is given by (compare (5.42) and (5.43))

$$\mathbf{M} = (\mathbf{I} - \mathbf{P}_{2,1}(\mathbf{N}_{s,1}^{-1} + \mathbf{P}_{1,0}\mathbf{S}_{s,0}^b \mathbf{R}_{0,1}(\mathbf{I}_1 - \mathbf{A}_1 \mathbf{N}_{s,1}^{-1}))\mathbf{R}_{1,2}\mathbf{A}_2)\mathbf{M}_{s,2}.$$

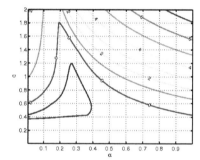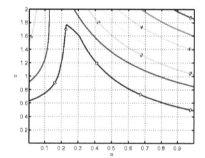

FIGURE 5.8: 2-stage method: Contourplots of $\log(\rho(\mathbf{M}(\alpha, c)))$ for $\Delta x = 1/24$ and $\nu = 0.125$ (left) and $\nu = 1.0$ (right).

To solve the optimization problem (5.48), we again use a grid search within a MATLAB code employing the eig function to obtain the spectral radius. For the 2-stage case, we use a grid of size $200 \times (2 \cdot 200)$ with $c \in [0, 2]$. Here, both ν and Δx have to be varied, where we chose ν according to the same CFL numbers as for the last strategy and $\Delta x = 1/24, 1/12$. The results are shown in Table 5.5. The contour lines of the function $\rho(\mathbf{M}(\alpha, c; \nu, \Delta x))$ are illustrated in Figure 5.8. As can be seen, the results are qualitatively similar to the ones from the other strategy.

In the 3-stage case, we use a grid of the size $100 \times 100 \times (2 \cdot 100)$. Finally, a $50 \times 50 \times 50 \times (2 \cdot 50)$ grid is used for the 4-stage case. The decrease in mesh width is due to the polynomially growing computational cost.

Comparing the results for the different stages, we see that there is no dependence of the optimal solution on Δx in the sense that for a fixed CFL

	$\Delta x = 1/24$					$\Delta x = 1/12$				
#stages	α_1	α_2	α_3	c	$\rho(\mathbf{M})$	α_1	α_2	α_3	c	$\rho(\mathbf{M})$
2	0.295			1.46	0.6425	0.29			1.485	0.6371
3	0.12	0.38		2.14	0.5252	0.10	0.31		2.42	0.4720
4	0.02	0.12	0.34	2.16	0.5173	0.02	0.1	0.30	2.18	0.5094

TABLE 5.5: Results of optimization of $\rho(\mathbf{M})$ for 2-stage scheme

number and accordingly chosen ν, the results are almost identical. In this sense, the multigrid method obtained has a convergence speed which is independent of the mesh width. Regarding the size of the spectral radius, going from two to three stages leads to a decrease of the optimal value. As for adding a fourth stage, this actually leads to an increase in spectral radius. This can be explained by the much coarser grid used for the optimization of the 4-stage method. Consequently, the solution found is too far away from the optimum to beat the 3-stage method.

In Figure 5.9, the stability region of the optimal methods for $\Delta x = 1/24$ and $\nu = 0.125$, as well as $\nu = 1.0$ are shown. Again, an increase in the number of stages leads to a larger stability region.

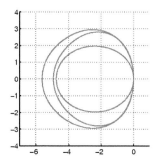 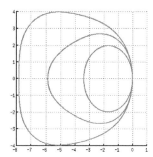

FIGURE 5.9: Stability region of optimal 2-, 3-, and, 4-stage method for $\Delta x = 1/24$ and $\nu = 0.125$ (left) and $\nu = 1.0$ (right). The larger the number of stages, the larger the stability region.

Comparing the optimal solutions found with the schemes obtained by the previous method, we see that the coefficients are similar, as is the value of c and the same is valid for the stability regions.

5.6.7.5 Numerical results

We test the smoothers on two problems with $\Delta x = 1/24$ on the finest level and $a = 2.0$. As initial conditions, we use a step function with values 5 and 1, as well as the function $\sin(\pi x)$. We then perform one time step with $\Delta t = 0.5$, which means $\nu = 1.0$ and $CFL = 24$. The resulting linear equation system is solved with 80 steps of the different multigrid methods. As a reference, the optimal 2- and 3-stage methods derived by van Leer et al. [287] for steady state problems are used. The 2-stage method is given by $\alpha = 1/3$ and $c = 1$, whereas the 3-stage method does not actually fall into the framework considered here. It consists of one step of the explicit Euler method with $c = 0.5$, followed by a step of a 2-stage method with $\alpha = 0.4$ and $c = 1$. All computations are performed using MATLAB.

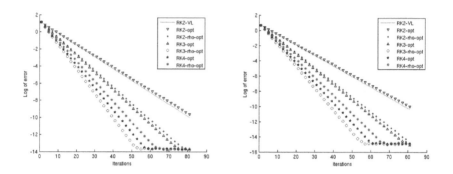

FIGURE 5.10: Convergence plots for different multigrid methods and $CFL = 24$: step function (left) and sine initial data (right).

Regarding computational effort, the main work of the multigrid method consists of matrix vector multiplications in the smoother. Thus, the 3-stage schemes are conservatively 50% more costly than the 2-stage schemes, whereas the 4-stage schemes are less than twice as expensive as the RK-2 smoother.

We now look at the convergence speed of the different methods, where we call the methods obtained by the second optimization ρ-optimized schemes. In Figure 5.10, $\log_{10}$ of the error in the 2-norm is plotted over multigrid iterations. The 3-stage method of van Leer diverges and is only shown in the first figure, where it is barely faster than the 2-stage method of van Leer. Otherwise we can see that the ρ-optimized schemes behave as expected where the 3-stage scheme is the fastest, followed by the 4-stage and the 2-stage schemes, with the 3-stage scheme being roughly twice as fast as the 2-stage scheme. For the schemes coming out of the first optimization, the 4-stage scheme is faster than the 3-stage scheme, which is faster than the 2-stage scheme. Furthermore, the ρ-optimized schemes are able to beat their counterparts with the exception of the 4-stage scheme. Thus, the more costly optimization is generally worthwhile.

Generally, the 3-stage ρ-optimized scheme is the fastest, in particular it is almost twice as fast as the 2-stage ρ-optimized scheme, making it more efficient. Compared to the reference method of van Leer, it is between two and four times faster, making it between 70% and 270% more efficient. Thus, just by changing the coefficients of the RK smoother, we can expect to gain more than a factor of 2 in multigrid efficiency.

5.6.7.6   Local Fourier analysis

A classic technique to analyze multigrid methods is local Fourier analysis (LFA) [281], which is often called discrete Fourier analysis (DFA) in the CFD community [119]. It follows the same concepts as the von Neumann stability analysis, except that there, the stability behavior for time stepping is considered. Here, we instead look at the convergence behavior of iterations, including smoothing factors. We now sketch the idea. It is a technique that can be used for multidimensional problems, but requires a discretized linear problem with periodic boundary conditions on a regular grid with possibly different mesh widths in the different coordinate directions and a linear iterative method applied to it. For nonlinear problems, this is obtained by linearizing in a point. Using lexicographic ordering, we associate the unknown vector $\mathbf{u}_{ij} \in \mathbb{R}^{(d+2)\times(d+2)}$ with cell ij. The method is thus a linear operator acting on a linear finite difference discretization and can be written in the absence of forcing terms as

$$\mathbf{G}\mathbf{u} = \mathbf{0}$$

with $\mathbf{G} \in \mathbb{R}^{m\times m}$.

The first step is to write the discretization in terms of (index) shift operators E_x and E_y, in x and y direction, respectively. This means that $E_x u_{i,j} = u_{i+1,j}$ and $E_y^{-1} u_{i,j} = u_{i,j-1}$. Thus,

$$u_{ij} - u_{i-1,j} = (1 - E_x^{-1})u_{ij}.$$

For the second ingredient we consider the solution as a function which has values in grid points. For periodic boundary conditions, we can then write it as a discrete Fourier series (here in two dimensions)

$$\mathbf{u}_{ij} = \sum_{k_x=-n_x/2+1}^{n_x/2} \sum_{k_y=-n_y/2+1}^{n_y/2} \hat{\mathbf{u}}_{k_x,k_y} e^{2\pi i(k_x x_i + k_y y_j)}$$

with the wavenumbers k_x, k_y, the number of unknowns n_x, n_y in both space directions, and $\hat{\mathbf{u}}_{k_x,k_y} \in \mathbb{R}^{(d+2)\times(d+2)}$. When applying a shift operator to one of the exponentials, we obtain the shifting equality

$$E_x e^{2\pi i(k_x x_i + k_y y_j)} = e^{2\pi i(k_x(x_i+1/n_x)+k_y y_j)} = e^{2\pi i k_x/n_x} e^{2\pi i(k_x x_i + k_y y_j)}$$

and similar for E_y. Thus, a discrete operator $\mathbf{G}$ acting on $\mathbf{u}$ has been (block) diagonalized and we can compute the spectral radius or the smoothing factor by computing a maximum over small matrices $\hat{\mathbf{G}} \in \mathbb{R}^{(d+2)\times(d+2)}$, called the Fourier symbols of $\mathbf{G}$, which act on the $\hat{\mathbf{u}}_{k_x,k_y}$.

To illustrate this, consider a second-order central finite difference of the pressure:

$$p_{i+1,j} - 2p_{ij} + p_{i-1,j} = (E_x^{+1} - 2 + E_x^{-1})p_{ij}.$$

In an analysis in conservative variables, we need to consider

$$p = (\gamma - 1) \left(\rho E - \rho \frac{(\rho v_1)^2 + (\rho v_2)^2}{2\rho^2} \right)$$

and what an application of shift operators does here. In the fraction, all shift operators cancel out. Thus, we get

$$p_{i+1,j} - 2p_{ij} + p_{i-1,j} = (\gamma-1)[(E_x^{+1} - 2 + E_x^{-1})\rho E_{ij} - |\mathbf{v}_j|^2/2(E_x^{+1} - 2 + E_x^{-1})\rho_{ij}].$$

Defining the phase angles

$$\Theta_x = 2\pi k_x/n_x, \quad \Theta_y = 2\pi k_y/n_y,$$

the Fourier symbols of the shift operators are

$$\hat{E}_x = e^{i\Theta_x}, \quad \hat{E}_y = e^{i\Theta_y}.$$

To compute the spectral radius of $\mathbf{G}$, we now just need to look at the maximum of the spectral radius of $\hat{\mathbf{G}}_{\Theta_x,\Theta_y} = \hat{\mathbf{G}}_{k_x,k_y}$ over all phase angles Θ_x and Θ_y between $-\pi$ and π. Furthermore, this allows to compute the smoothing factor as well, by instead taking the maximum over half the wave number range.

In the example, we obtain

$$\hat{G}(\Theta)p = (\gamma - 1)[(e^{i\Theta} - 2 + e^{-i\Theta})\rho E - |\mathbf{v}|^2/2(e^{i\Theta} - 2 + e^{-i\Theta})\rho.$$

For a given linearization point $(\rho, \rho\mathbf{v}, \rho E)$, this can be computed.

One example where this technique has been applied to discretizations for compressible flows is [30], as well as [158].

5.6.7.7   Generalized locally Toeplitz sequences

A related analysis technique is that of generalized locally Toeplitz (GLT) sequences [97]. We consider a PDE on the spatial interval $[0, 1]$. We start with the notion of a symbol of a Toeplitz matrix: Given a function $f \in L_1(-\pi, \pi)$, the symbol, we can define the Toeplitz matrix $T_m(f) = (\hat{f}_{i-j})_{i,j} \in \mathbb{R}^{m \times m}$ with the Fourier coefficient

$$\hat{f}_k = \frac{1}{2\pi} \int_\pi^\pi f(\Theta)e^{-ik\Theta}d\Theta.$$

As an example, $f(x) = 1 - e^{-ix}$ gives the matrix corresponding to the upwind stencil for $u_t + au_x = 0$ with periodic boundary conditions. There is a rich literature about the relation between the symbol and the spectrum of Toeplitz matrices, in particular asymptotic results for $m \to \infty$ for normal matrices. We will not give a review, because we are interested in nonnormal block matrices anyhow, which are also not Toeplitz matrices. However, statements about the asymptotic distribution of singular values can still be made using the concept of GLT sequences.

Any GLT sequence $\{\mathbf{A}_m\}_m$ has a symbol κ and we write

$$\{\mathbf{A}_m\}_m \sim_{GLT} \kappa.$$

We have:

$$\{\mathbf{A}_m\}_m \sim_{GLT} \kappa \Rightarrow \{\mathbf{A}_m\}_m \sim_\sigma \kappa.$$

Here, $\{\mathbf{A}_m\}_m \sim_\sigma \kappa$ denotes that the singular values of the matrix sequence $\{\mathbf{A}_m\}_m$ are asymptotically distributed as κ, which means that

$$\lim_{m\to\infty} \frac{1}{m} \sum_{i=1}^m F(\sigma_i(\mathbf{A}_m)) = \frac{1}{\mu_k(D)} \int_D F(|\kappa(x,\Theta)|)dxd\Theta \quad \forall F \in C_c(D),$$

with $D = [0,1] \times [-\pi, -\pi]$. Thus, the singular values are asymptotically well approximated by sampling the symbol in equidistant points.

This concept is useful for several reasons. The first is that GLT sequences form an algebra, which means that if $\{\mathbf{A}_m\}_m$ and $\{\mathbf{B}_m\}_m$ are GLT sequences with symbols κ and ξ, then for complex numbers α and β

$$\alpha\{\mathbf{A}_m\}_m + \beta\{\mathbf{B}_n\}_n \sim_{GLT} \alpha\kappa + \beta\xi,$$

$$\{\mathbf{A}_m\}_m\{\mathbf{B}_m\}_m \sim_{GLT} \kappa\xi.$$

Furthermore, the Toeplitz matrix sequence $\{\mathbf{T}_m(f)\}_m$ is a GLT sequence with symbol $\kappa(x,\Theta) = f(\Theta)$. Additionally, diagonal sampling matrices where the diagonal entries do a uniform sampling of a continuous function $a(x)$ form GLT sequences with symbol $\kappa(x,\Theta) = a(x)$. These properties together allow to find GLT sequences and to compute their spectra.

The reason why this concept is interesting for compressible flow problems is that it can be extended to matrices that are not Toeplitz matrices in one additional way. It can be proved that matrix sequences that are block Toeplitz matrices plus a low rank and a low norm perturbation are GLT sequences.

Overall, nonperiodic boundary conditions, nonequidistant meshes and variable coefficients, i.e., nonlinear problems, can be treated. This technique has been successfully applied to analyze multigrid methods for discretizations for convection-dominated flows in [21] and to higher order methods for diffusion-dominated flows in [77].

5.7 Newton's method

A classical method to solve nonlinear equation systems is Newton's method, sometimes referred to as the Newton-Raphson method. The basic idea is to linearize the problem and thus replace the nonlinear problem by a sequence of linear problems. For the one-dimensional case, the method is illustrated in Figure 5.11. Newton's method is locally convergent and can exhibit quadratic convergence. Outside the region of convergence, it is not clear what happens. The method can diverge or it converges to a solution of equation (5.52). This is problematic if there are multiple solutions. In fact, it has been shown for a number of equations that the set of starting points from which the method converges to a certain solution is fractal.

The local convergence means that its use for steady flows is limited, since the typical starting guess in that case consists of choosing freestream values in the whole domain, which are far away from the steady state values. This problem can be addressed by globalization strategies that are linearly convergent in an initial phase, thus often being slow. For unsteady problems, the situation is different where the solution at time t_n is not that far away from the solution at t_{n+1}. Therefore, Newton's method has to be reassessed in that situation. Jothiprasad et al. compare the FAS scheme with an inexact Newton method where either linear multigrid or GMRES is used to solve the linear systems and find the FAS scheme to be computationally the worst [149].

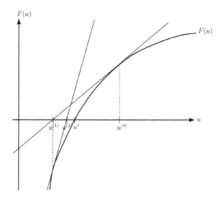

FIGURE 5.11: Illustration of Newton's method in one dimension. (Credit: V. Straub.)

We now explain those parts of the theory relevant to the methodology used here and otherwise refer to the books of Kelley [154] and Deuflhard [76]. Newton's method solves the root problem

$$\mathbf{F}(\underline{u}) = \underline{0} \tag{5.52}$$

for a differentiable function $\mathbf{F}(\underline{u})$ in the following way:

$$\left.\frac{\partial \mathbf{F}(\underline{u})}{\partial \underline{u}}\right|_{\underline{u}^{(k)}} \Delta \underline{u} = -\mathbf{F}(\underline{u}^{(k)}) \tag{5.53}$$

$$\underline{u}^{(k+1)} = \underline{u}^{(k)} + \Delta \underline{u}, \quad k = 0, 1, \dots.$$

The method must not necessarily be written in conservative variables. Instead, derivatives of primitive variables could be used without any difference in convergence speed. As termination criteria, either (5.8) or (5.9) is used.

Convergence of the method can be proved under very few assumptions. However, a major point about Newton's method is that it is second-order convergent. To prove that, the following standard assumptions are used in [154]:

Definition 8 (Standard Assumptions)

i) Equation (5.52) has a solution $\underline{u}^*$.

ii) $\underline{F}' : \Omega \to \mathbb{R}^{m \times m}$ is Lipschitz continuous with Lipschitz constant L'.

iii) $\underline{F}'(\underline{u}^*)$ is nonsingular.

Note: Since the determinant is a continuous function, (iii) implies with (ii) that F' is nonsingular in a whole neighborhood around x^* and thus, Newton's method is well defined in that neighborhood.

If the linear equation systems (5.53) are solved exactly, the method is locally second-order convergent. This is never done, because an exact Jacobian is rarely available and furthermore, this is too costly and unnecessary. Instead, we will approximate terms in (5.53) and solve the linear systems only approximately. Thus we might lose convergence speed, but we save a significant amount of CPU time. Since in case of convergence, the limit is determined by the right-hand side, we can approximate the matrix or $\Delta \underline{u}$ without disturbing the outer time integration scheme.

5.7.1 Simplified Newton's method

If we approximate the matrix, there is first of all the simplified Newton's method, where we compute the Jacobian only once and thus iterate:

$$\left. \frac{\partial \mathbf{F}(\underline{u})}{\partial \underline{u}} \right|_{\underline{u}^{(0)}} \Delta \underline{u} = -\mathbf{F}(\underline{u}^{(k)}) \tag{5.54}$$

$$\underline{u}^{(k+1)} = \underline{u}^{(k)} + \Delta \underline{u}, \quad k = 0, 1, \dots.$$

This method is also locally convergent and has convergence order 1. It is frequently used in a variant, where the Jacobian is recomputed periodically, see, for example, [196]. Due to the high cost of computing the Jacobian, the loss of order of convergence is typically less important.

5.7.2 Methods of Newton type

The simplified Newton method can be generalized to so-called methods of Newton type, where we approximate the Jacobian by some matrix $\mathbf{A}$. For example, this could correspond to a lower order Jacobian for a higher order discretization. We thus obtain the scheme:

$$
\begin{aligned}
\mathbf{A}\Delta\underline{\mathbf{u}} &= -\mathbf{F}(\underline{\mathbf{u}}^{(k)}) \\
\underline{\mathbf{u}}^{(k+1)} &= \underline{\mathbf{u}}^{(k)} + \Delta\underline{\mathbf{u}}, \quad k = 0, 1, ...
\end{aligned}
\tag{5.55}
$$

This method converges locally linear if $\mathbf{A}$ is close enough to the Jacobian or more precise, if $\rho(\mathbf{I} - \mathbf{A}^{-1}\frac{\partial\mathbf{F}(\underline{\mathbf{u}})}{\partial\underline{\mathbf{u}}}) < 1$, respectively if $\|\mathbf{A} - \frac{\partial\mathbf{F}(\underline{\mathbf{u}})}{\partial\underline{\mathbf{u}}}\|$ is small enough. The size of the neighborhood of the solution where convergence can be proven depends on the spectral radius just mentioned where it becomes smaller, the further away $\mathbf{A}$ is from $\frac{\partial\mathbf{F}(\underline{\mathbf{u}})}{\partial\underline{\mathbf{u}}}$.

5.7.3   Inexact Newton methods

Another approach to saving computing time are inexact Newton methods. There, the linear equation systems are solved by an iterative scheme, for example one of the later discussed Krylov subspace methods. These are terminated prematurely, based on the residual of the linear equation system. Typically, the scheme suggested in [74] is chosen, where the inner solver is terminated, if the relative residual is below a certain threshold. This type of scheme can be written as

$$
\left\|\frac{\partial\mathbf{F}(\underline{\mathbf{u}})}{\partial\underline{\mathbf{u}}}\Big|_{\underline{\mathbf{u}}^{(k)}}\Delta\underline{\mathbf{u}} + \mathbf{F}(\underline{\mathbf{u}}^{(k)})\right\| \leq \eta_k\|\mathbf{F}(\underline{\mathbf{u}}_k)\|
\tag{5.56}
$$

$$
\underline{\mathbf{u}}^{(k+1)} = \underline{\mathbf{u}}^{(k)} + \Delta\underline{\mathbf{u}}, \quad k = 0, 1, ...
$$

The $\eta_k \in \mathbb{R}$ are called forcing terms. In [83], the choice for this sequence is discussed and the following theorem is proved.

Theorem 7 *Let the standard assumptions hold. Then there is δ such that if $\underline{\mathbf{u}}^{(0)}$ is in a δ-neighborhood of $\underline{\mathbf{u}}^*$, $\{\eta_k\} \subset [0, \eta]$ with $\eta < \bar{\eta} < 1$, then the inexact Newton iteration (5.56) converges linearly. Moreover,*

- *if $\eta_k \to 0$, the convergence is superlinear and*

- *if $\eta_k \leq K_\eta\|\mathbf{F}(\underline{\mathbf{u}}^{(k)})\|^p$ for some $K_\eta > 0$ and $p \in [0, 1]$, the convergence is superlinear with order $1 + p$.*

The region of convergence given by δ is smaller or as large as the one for the exact Newton's method. Furthermore, the convergence region decreases with growth of $\kappa(\mathbf{F}'(\mathbf{u}))$. Regarding speed of convergence, it follows from the theorem that the common strategy of choosing a fixed η leads to a first-order convergence speed. The last part of the theorem means that for a properly chosen sequence of forcing terms, namely, if the η_k converge to zero fast enough, convergence is quadratic. However, it is not necessary to solve the first few linear systems very accurately. This is in line with the intuition that while we are far away from the solution, we do not need the optimal search direction for Newton's method, but just a reasonable one to get us in the generally correct direction.

A way of achieving this is the following:

$$\eta_k^A = \gamma \frac{\|\mathbf{F}(\underline{\mathbf{u}}^{(k)})\|^2}{\|\mathbf{F}(\underline{\mathbf{u}}^{(k-1)})\|^2}$$

with a parameter $\gamma \in (0, 1]$. This was suggested by Eisenstat and Walker in [83], where they also prove that this sequence has the convergence behavior required for the theorem. Thus, the theorem says that if this sequence is bounded away from one uniformly, convergence is quadratic. Therefore, we set $\eta_0 = \eta_{\max}$ for some $\eta_{\max} < 1$ and for $k > 0$:

$$\eta_k^B = \min(\eta_{\max}, \eta_k^A).$$

Eisenstat and Walker furthermore suggest safeguards to avoid volatile decreases in η_k. To this end, $\gamma \eta_{k-1}^2 > 0.1$ is used as a condition to determine if η_{k-1} is rather large and thus the definition of η_k is refined to

$$\eta_k^C = \begin{cases} \eta_{\max}, & n = 0, \\ \min(\eta_{\max}, \eta_k^A), & n > 0, \gamma \eta_{k-1}^2 \leq 0.1 \\ \min(\eta_{\max}, \max(\eta_k^A, \gamma \eta_{k-1}^2)) & n > 0, \gamma \eta_{k-1}^2 > 0.1 \end{cases}$$

Finally, to avoid oversolving in the final stages, Eisenstat and Walker suggest

$$\eta_k = \min(\eta_{\max}, \max(\eta_k^C, 0.5\tau/\|\mathbf{F}(\underline{\mathbf{u}}^{(k)})\|)), \tag{5.57}$$

where τ is the tolerance at which the Newton iteration would terminate (see (5.8) and (5.9)).

5.7.4 Choice of initial guess

The choice of the initial guess in Newton's method is important for two reasons: Firstly, the method is only locally convergent, meaning that for bad choices of initial guesses, the method may not converge. Secondly, a smart initial guess may reduce the number of iterations significantly.

A reasonable starting value for the Newton iterations for unsteady computations is the solution from the last time level and the last stage value inside a DIRK scheme, respectively. For small time steps, this will be in the region of convergence; if the time step is too large, the method will diverge. Typically, this type of initial guess works for reasonably large time steps. If not, as discussed in section 5.4, we will repeat the time step with half the step size, hopefully leading to a situation where the initial guess is in the region of convergence. Nevertheless, better initial guesses can be obtained, for example by extrapolation, proper orthogonal decomposition [280], or dense output formulas (4.24). Extrapolation corresponds to taking the sequence of solutions of the previous nonlinear systems and extrapolating it in some way to the next system. Dense output formulas originate in the DIRK methods used.

5.7.5   Globally convergent Newton methods

Even with an improved initial guess obtained using any of the methods from the last section, we might not be in the region of convergence of Newton's method. This makes globally convergent variants of Newton's method attractive. This property can be achieved in a number of ways. Here, we focus on using a so-called line search, which is easy to implement. This means that we change the update in the method to

$$\underline{u}^{(k+1)} = \underline{u}^{(k)} + \lambda \Delta \underline{u}^{(k)},$$

where λ is a real number between 0 and 1. Different ways of doing this line search exist, a popular choice is the Armijo line search [75]. Here, λ is chosen such that we obtain a sequence of residuals that is almost monotone:

$$\|\mathbf{f}(\underline{u}^{(k+1)})\| < (1 - \alpha\lambda)\|\mathbf{f}(\underline{u}^{(k)})\|, \tag{5.58}$$

with α being a small positive number. For example, $\alpha = 10^{-4}$ is a suitable choice. The search for a suitable λ is done in a trivial way by starting with 1 and then dividing it by 2 if the condition (5.58) is not satisfied. More sophisticated strategies are possible.

It can be shown that this method is linearly convergent [75]; however, once the iteration is close enough to the solution, full steps with $\lambda = 1$ can be taken and thus, the original, possibly quadratic, convergence behavior is recovered.

Globally convergent Newton methods can be used to compute steady state solutions. However, this does not lead to schemes that are competitive with the FAS multigrid. A variant that is often used is to consider the backward Euler method with just one Newton step per time step and then choose a CFL number that increases with time. This corresponds to a damped Newton method, where the choice of the λ from above does not originate in a line search, but a time step chosen. In Figure 5.12, the norm of the steady state residual is shown for a run of TAU_2D (see chapter 7) for a NACA0012 profile computation. The Mach number is 0.85, the Reynolds number is 1.000, the angle of attack 1.25, the computation is started with freestream values and the number of cells is 4605.

The starting CFL number is 4 and this is increased every 10 steps by 1.2. While this is not an optimal strategy, it works reasonably well. The Newton termination tolerance is 10^{-4}, as is the case for the linear systems. As can be seen, convergence to the steady state is very slow, which explains why Newton methods are rarely used for steady state computations.

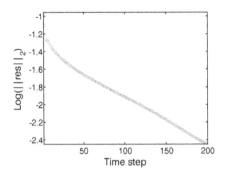

FIGURE 5.12: Residual history for a damped Newton method for the computation of the steady state around a NACA0012 profile.

5.7.6 Computation and storage of the Jacobian

The computation of the Jacobian leads to two significant problems. First, as explained in the introduction of this chapter, the function $\mathbf{F}$ is in general only piecewise differentiable and thus, the Jacobian in $\mathbf{u}^{(k)}$ may not exist. Second, if it exists, the computation of the full second-order Jacobian in the finite volume context is extremely difficult to implement and to compute. In the DG context, the implementation and computation is much easier, but in particular the viscous fluxes pose a certain amount of difficulty. Therefore, it is important to consider ways of circumventing these two problems.

First of all, approximations of the Jacobian can be used. For example, we could compute a Jacobian based on the first-order method, which does not contain a reconstruction procedure or limiters and can therefore be implemented and computed in a reasonable amount of time. This means that a method of Newton type (5.55) is employed and thus only first-order convergence is reached. Nevertheless, this is a popular approach.

Alternatively, we could use automatic differentiation. This is a procedure where the code itself is analyzed by a program (for example ADIFOR [42] or Tapenade [127]), which automatically creates a second version of the code, that computes the derivative using product rule, quotient rule, and so on. If-statements are treated as branches of the Jacobian definition.

Both of these approaches have the drawback that the Jacobian needs to be stored. Since it is sparse, this is done using a sparse data format, for example compressed row storage (CRS). There, two vectors are stored for each row, one integer vector for the indices of columns where nonzero entries can be found and a second one of equal length with the corresponding values. The block structure can be respected by letting the indices correspond to cells and changing the second vector from one with values to one with pointers, which point to an appropriate container for values of the small block. The number of nonzeros for three-dimensional flows is 35 per unknown for quadrilaterals and 25 per unknown for tetrahedrons for finite volume schemes, whereas the number explodes for DG methods, leading to an often prohibitive amount of storage. Therefore, often so-called Jacobian-free methods will be used in conjunction with Krylov-subspace methods. This is explained in the next section.

5.8   Krylov subspace methods

As mentioned in section 5.2, we use Krylov subspace methods for the linear system [241, 284, 249]. These approximate the solution of a linear system

$$\mathbf{A}\mathbf{x} = \mathbf{b}, \quad \mathbf{A} \in \mathbb{R}^{m \times m}$$

in the space

$$\mathbf{x}_0 + \text{span}\{\mathbf{r}_0, \mathbf{A}\mathbf{r}_0, ..., \mathbf{A}^{m-1}\mathbf{r}_0\} = \mathbf{x}_0 + \mathcal{K}_m(\mathbf{A}, \mathbf{r}_0) \tag{5.59}$$

with $\mathbf{x}_0$ being the initial guess. The space $\mathcal{K}_m$ is called the mth Krylov subspace of $\mathbf{A}, \mathbf{b}$, and $\mathbf{x}_0$. Thus, the approximate solution in step m is of the form $\mathbf{x}_m = \mathbf{x}_0 + p(\mathbf{A})\mathbf{r}_0$ with p being a polynomial of degree $m - 1$. A relevant special case for our problems is that the initial guess is zero. Then, $\mathbf{r}_0 = \mathbf{b}$ and the search space is span$\{\mathbf{b}, \mathbf{A}\mathbf{b}, ..., \mathbf{A}^{m-1}\mathbf{b}\}$.

A prototype Krylov subspace method computes an orthonormal basis of this space and then uses a projection to compute the next iterate. This process needs the Jacobian only for matrix vector multiplications and thus the sparsity of $\mathbf{A}$ can be exploited.

In the case of symmetric and positive definite systems, the conjugate gradient (CG) method computes the iterate with optimal error in the energy norm $\sqrt{\mathbf{x}^T\mathbf{A}\mathbf{x}}$ using a short recurrence of three vectors. For compressible flows, the matrices will be unsymmetric, thus CG cannot be used. As Faber and Manteuffel proved in [87], a similar method that is both optimal and has a short recurrence is not possible for general matrices. Therefore, a choice has to be made: Either use a method that is optimal and thus has increasing demand for storage and CPU time or use a method that has a short recurrence and thus a constant storage and CPU time requirement per iteration or use a method like CGNR that applies CG to the normal equations $\mathbf{A}^T\mathbf{A}\mathbf{x} = \mathbf{A}^T\mathbf{b}$ and thus the matrix $\mathbf{A}^T\mathbf{A}$. This comes at the price of a squaring of the condition number and the use of $\mathbf{A}^T$. As we will explain later, we prefer not to store the Jacobian and thus do not expect to have access to its transpose. Thus we do not suggest the use of these schemes.

The most prominent methods of the first two kinds are GMRES of Saad and Schultz [242], which computes the optimal iterate regarding the 2-norm of the residual, but does not have a short recurrence, and BiCGSTAB of van der Vorst [283], which has a three-vector recurrence, but does not satisfy an optimality property. Furthermore, the interesting method IDR(s) has been recently introduced by van Gijzen and Sonneveld, which is a generalization of BiCGSTAB [256]. In [209], different Krylov subspace methods are applied to different classes of nonsymmetric systems and for each type, a class is found where this class performs best. Therefore, no general answer can be given, as to which method performs best for nonsymmetric matrices. The main reason for this is that within the class of nonsymmetric matrices we have the nonnormal ones and there, the eigenvalues and eigenvectors no longer tell us everything about the properties of the matrix. A description of the current state of theory is given in [200].

5.8.1 GMRES and related methods

As a first Krylov subspace method we will explain the generalized minimal residual method (GMRES) [242] in detail. In the jth iteration, the scheme computes an orthonormal basis $\mathbf{v}_1, ..., \mathbf{v}_j$ of the jth Krylov subspace

$$\mathcal{K}_j = \mathrm{span}\{\mathbf{r}_0, ..., \mathbf{A}^{j-1}\mathbf{r}_0\}$$

by the so-called Arnoldi method. GMRES uses this basis to minimize the functional

$$J(\mathbf{x}) = \|\mathbf{A}\mathbf{x} - \mathbf{b}\|_2$$

in the space $\mathbf{x}_0 + \mathcal{K}_j$. The point about Arnoldi's method is that it allows an efficient implementation using a Hessenberg matrix. In every step, this is updated and transformed to an upper triangular matrix using for example Givens-rotations. This then allows to obtain the value of the functional J in every iteration without explicit computation of $\mathbf{x}_j$, which is only done after the tolerance is satisfied. In pseudocode, the algorithm can be formulated as (taken from [197]):

- $\mathbf{r}_0 = \mathbf{b} - \mathbf{A}\mathbf{x}_0$

- If $\mathbf{r}_0 = \mathbf{0}$, then END

- $\gamma_1 = \|\mathbf{r}_0\|$

- $\mathbf{v}_1 = \frac{\mathbf{r}_0}{\|\mathbf{r}_0\|_2}$

- For $j = 1, ..., m$

  - $\mathbf{w}_j = \mathbf{A}\mathbf{v}_j$
  - For $i = 1, ..., j$ do $h_{ij} = \mathbf{v}_i^T \mathbf{w}_j$
  - $\mathbf{w}_j = \mathbf{w}_j - \sum_{i=1}^{j} h_{ij}\mathbf{v}_i, \quad h_{j+1,j} = \|\mathbf{w}_j\|_2$
  - For $i = 1, ..., j-1$ do $\begin{pmatrix} h_{ij} \\ h_{i+1,j} \end{pmatrix} = \begin{pmatrix} c_{i+1} & s_{i+1} \\ -s_{i+1} & c_{i+1} \end{pmatrix} \begin{pmatrix} h_{ij} \\ h_{i+1,j} \end{pmatrix}$

- $\beta = \sqrt{h_{jj}^2 + h_{j+1,j}^2}; \quad s_{j+1} = \frac{h_{j+1,j}}{\beta}$

- $c_{j+1} = \frac{h_{jj}}{\beta}; \quad h_{jj} = \beta.$

- $\gamma_{j+1} = -s_{j+1}\gamma_j; \quad \gamma_j = c_{j+1}\gamma_j$

- if $|\gamma_{j+1}| \geq TOL$, $\mathbf{v}_{j+1} = \frac{\mathbf{w}_j}{h_{j+1,j}}$

- else

  * for $i = j, ..., 1$ do $\alpha_i = \frac{1}{h_{ii}}\left(\gamma_i - \sum_{k=i+1}^{j} h_{ik}\alpha_k\right)$

  * $\mathbf{x} = \mathbf{x}_0 + \sum_{i=1}^{j} \alpha_i \mathbf{v}_i$

  * END

Since GMRES has no short recurrence, the whole basis has to be stored. Thus the cost and storage per step increase linearly with the iteration number. In step j, we need to store the j basis vectors and compute one matrix vector product and j scalar products. If only few iterations are needed, this is dominated by the matrix vector product, which is cheap. Thus, GMRES works very well in inexact Newton methods using the Eisenstat-Walker strategy, since in the first Newton steps, only one or two GMRES iterations are needed.

It is possible to restart the iteration after k iterations by scrapping the orthogonal basis and starting new with the current approximation, thus bounding the maximal amount of storage without simply terminating the iteration. This method is called GMRES(k). Unfortunately, there are examples where the restart technique does not lead to convergence and examples, where the restart convergence results in a speedup. Typical restart lengths are 30 or 40.

The assumption used in the termination criteria is that the basis is actually orthonormal. This can be tested by computing the residual after the computation of the solution. It turns out that in practice, the difference in the estimate and the actual residual is negligible. Therefore, we will never use techniques for reorthonormalization of the basis used in GMRES, as suggested by several authors.

Due to the minimization, in exact arithmetic, GMRES computes the exact solution of an $m \times m$ linear equation system in at most m steps. Furthermore, the residual in the 2-norm is nonincreasing in every step. For diagonalizable matrices, a more useful residual estimate is the following (for the proof of this and the next results, see e.g. [154]):

Theorem 8 *Let* $\mathbf{A} = \mathbf{V}\mathbf{\Lambda}\mathbf{V}^{-1}$ *be a nonsingular diagonalizable matrix. Then for all* $p \in \Pi_k$ *with* $p(0) = 1$, *we have*

$$\|\mathbf{r}_{n+1}\|_2 \leq \|\mathbf{r}_0\|_2 \kappa_2(\mathbf{V}) \max_{z \in \sigma(\mathbf{A})} |p_k(z)|.$$

Here, $\sigma(\mathbf{A})$ denotes the spectrum of $\mathbf{A}$. It follows from this that if $\mathbf{A}$ is normal and the right-hand side has only m eigenvector components, GMRES will produce the solution in m iterations. The theorem also suggests the method will converge fast if the eigenvalues are clustered, since then the method will choose the polynomial such that the zeros are the eigenvalues. This causes a problem for us. Differential operators are unbounded and multiscale, which implies that the spectrum of a discrete operator will not be clustered for any reasonable discretization. Incidentally, this is the same reason that we obtain stiff problems. Thus, GMRES (and other Krylov subspace methods as well) will not be a good method for the problems considered here. This is why preconditioning, which is explained in the next chapter, is absolutely crucial.

For nonnormal matrices, the condition number is not a good tool to describe the behavior of the matrix $\mathbf{A}$. More useful is the distribution of eigenvalues. A result that sheds some light on the performance of GMRES for general matrices is the following:

Theorem 9 *Let $\mathbf{A}$ be nonsingular, then we have*

$$\mathbf{r}_k = \min_{p \in \Pi_k, p(0)=1} \| p(\mathbf{A})\mathbf{r}_0 \|_2.$$

However, if a matrix is strongly nonnormal, the eigenvalues can fail to accurately reflect the behavior of a matrix. Consequently, a theorem by Greenbaum, Pták, and Strakoš [113] states that for any discrete monotone decreasing function, a matrix with an arbitrary eigenvalue distribution can be constructed for which the GMRES algorithm produces that residual history. In particular it might be possible that the residual is constant until the very last step when it drops to zero. A concept that sheds more light on the convergence behavior for nonnormal matrices are the pseudospectra introduced by Trefethen [279]. Unfortunately, for the matrix dimensions considered here, these cannot be computed in reasonable time.

5.8.1.1   GCR

A method that is mathematically equivalent to GMRES is the generalized conjugate residual method (GCR) [82]. This uses two sets of basis vectors and thus allows some more flexibility in convergence acceleration, which will be discussed later.

- $\mathbf{x}_0 = \mathbf{0}$, $\mathbf{r}_0 = \mathbf{b} - \mathbf{A}\mathbf{x}_0$, k=-1

- while $\|\mathbf{r}_k\|_2 > tol$ do

    - $k = k + 1$
    - $\mathbf{p}_k = \mathbf{r}_k$
    - $\mathbf{q}_k = \mathbf{A}\mathbf{p}_k$
    - for $i = 0, 1, \ldots, k-1$ do $\alpha_i = \mathbf{q}_i^T \mathbf{q}_i$, $\mathbf{q}_k = \mathbf{q}_k - \alpha_i \mathbf{q}_i$, $\mathbf{p}_k = \mathbf{p}_k - \alpha_i \mathbf{p}_i$
    - $\mathbf{q}_k = \mathbf{q}_k / \|\mathbf{q}_k\|_2$, $\mathbf{p}_k = \mathbf{p}_k / \|\mathbf{q}_k\|_2$
    - $\mathbf{x}_{k+1} = \mathbf{x}_k + \mathbf{p}_k \mathbf{q}_k^T \mathbf{r}_k$
    - $\mathbf{r}_{k+1} = \mathbf{r}_k - \mathbf{q}_k \mathbf{q}_k^T \mathbf{r}_k$

5.8.2   BiCGSTAB

As the major alternative to the GMRES method, the biconjugate gradient stabilized (BiCGSTAB) method was established [283]. This uses two matrix vector products per iteration, but a constant number of basis vectors and scalar products. The method is constructed to satisfy

$$\mathbf{r}_k = q_k(\mathbf{A})p_k(\mathbf{A})\mathbf{r}_0$$

where $q_k, p_k \in \Pi_{k-1}$. In pseudocode, the method can be written as follows:

- $\mathbf{r}_0 = \mathbf{b}_0 - \mathbf{A}\mathbf{x}_0$. Set $\rho_0 = \alpha = \omega = 1$

- If $\mathbf{r}_0 = \mathbf{0}$, then END

- For $j = 1, ..., n$

  - $\beta = (\rho_k/\rho_{k-1})(\alpha/\omega)$
  - $\mathbf{p} = \mathbf{r} + \beta(\mathbf{p} - \omega\mathbf{v})$
  - $\mathbf{v} = \mathbf{A}\mathbf{p}$
  - $\alpha = \rho_k/(\hat{\mathbf{r}}_o^T\mathbf{v})$
  - $\mathbf{s} = \mathbf{r} - \alpha\mathbf{v}$, $\mathbf{t} = \mathbf{A}\mathbf{s}$
  - $\omega = \mathbf{t}^t\mathbf{s}/\|\mathbf{t}\|_2^2$, $\rho_{k+1} = -\omega\hat{\mathbf{r}}_0^T\mathbf{t}$
  - $\mathbf{x} = \mathbf{x} + \alpha\mathbf{p} + \omega\mathbf{s}$
  - $\mathbf{r} = \mathbf{s} - \omega\mathbf{t}$
  - END

For BiCGSTAB, no convergence analysis similar to the one of GMRES is known and in fact, the method can break down. What can be proven is that the residual after k steps of BiCGSTAB is never smaller than that after $2k$ steps of GMRES when starting with the same initial guess. Nevertheless, it exhibits fast convergence behavior in practice.

5.9   Jacobian-free Newton-Krylov methods

In Krylov subspace methods, the system matrix appears only in matrix vector products. Thus it is possible to formulate a Jacobian-free version of Newton's method. See [160] for a survey on Jacobian-free Newton-Krylov methods with a lot of useful references. To this end, the matrix vector products $\mathbf{A}\mathbf{q}$ are replaced by a difference quotient via

$$\mathbf{A}\mathbf{q} = \frac{\partial \mathbf{F}(\bar{\mathbf{u}})}{\partial \underline{\mathbf{u}}}\underline{\mathbf{q}} \approx \frac{\mathbf{F}(\bar{\mathbf{u}} + \epsilon\underline{\mathbf{q}}) - \mathbf{F}(\bar{\mathbf{u}})}{\epsilon}. \tag{5.60}$$

This works for the linear systems arising in Newton scheme, but also for those from the Rosenbrock scheme. For the root equation arising from (5.1), the above approximation translates into

$$\mathbf{A}(\bar{\mathbf{u}})\underline{\mathbf{q}} \approx \underline{\mathbf{q}} + \alpha \Delta t \left(\frac{\mathbf{f}(\bar{\mathbf{u}} + \epsilon\underline{\mathbf{q}}) - \mathbf{f}(\bar{\mathbf{u}})}{\epsilon} \right).$$

If the parameter ϵ is chosen very small, the approximation becomes better; however, cancellation errors become a major problem. A simple choice for the parameter that avoids cancellation but still is moderately small is given by Quin, Ludlow, and Shaw [226] as

$$\epsilon = \frac{\sqrt{eps}}{\|\underline{\mathbf{q}}\|_2},$$

where *eps* is the machine accuracy.

The Jacobian-free version has several advantages, the most important are low storage and ease of implementation: instead of computing the Jacobian by analytical formulas or difference quotients, we only need flux evaluations. In the context of finite volume solvers, the most important part of the spatial discretization are the approximate Riemann solvers. Changing one of these thus becomes much more simple. Furthermore, this allows to increase the convergence speed of the Newton scheme: Since we never compute the exact Jacobian, but only a first-order version, the scheme with matrix is always a method of Newton-type (5.55) and thus has first-order convergence. However, the Jacobian-free approximation is for the second-order Jacobian and thus can obtain second-order convergence if proper forcing terms are employed, since it is possible to view the errors coming from the finite difference approximation as arising from inexact solves and not from approximation to the Jacobian.

Of the Krylov subspace methods suitable for the solution of unsymmetric linear equation systems, the GMRES method was explained by McHugh and Knoll [193] to perform better than others in the Jacobian-free context. The reason for this is that the vectors in matrix vector multiplications in GMRES are normalized, as opposed to those in other methods.

Regarding convergence, there is the following theorem about the outer iteration in the JFNK scheme, when the linear systems are solved using GMRES (see [154]). Essentially, the Jacobian-free approximation adds another error of the order $\mathcal{O}(\epsilon)$. This can be interpreted as an increase of the tolerance η_k in the kth step to $\eta_k + c\epsilon$ for some constant c. Then, the previous theorem 7 can be applied.

Theorem 10 *Let the standard assumptions hold. Then there are δ, $\bar{\sigma}$, c such that if $\underline{\mathbf{u}}_0$ is in a δ-neighborhood of $\underline{\mathbf{u}}^*$ and the sequences $\{\eta_k\}$ and $\{\epsilon_k\}$ satisfy*

$$\sigma_k = \eta_k + c\epsilon_k \leq \bar{\sigma},$$

then the Jacobian-free Newton-GMRES iteration (5.56) converges linearly. Moreover,

- *if $\sigma_k \to 0$, the convergence is superlinear and*

- *if $\sigma_k \leq K_\eta \|\mathbf{F}(\underline{u}^{(k)})\|^p$ for some $K_\eta > 0$ and $p \in [0, 1]$, the convergence is superlinear with order $1 + p$.*

This theorem says that for superlinear convergence, the parameter ϵ in (5.60) needs to approach zero. However, this leads to cancellation errors. Therefore, if we keep ϵ fixed as described, a behavior like quadratic convergence can only be expected for the initial stages of the iterations, while $c\epsilon$ is still small compared to η_k. The constant c is problem dependent and thus it can happen that linear convergence sets in at a very late stage. In these cases, the strategy of Eisenstat and Walker for the choice of η_k still makes sense.

5.10   Comparison of GMRES and BiCGSTAB

We now compare GMRES, GMRES(25), and BiCGSTAB for the case with matrix and the Jacobian-free variant. In the following, we use the finite volume code TAU_2D (see chapter 7). We consider the linear system from an implicit Euler step with a certain CFL number, solve that up to machine accuracy, and look at the residual norm $\|\mathbf{A}x^{(k)} - \mathbf{b}\|_2$ over iterations. The first test problem is the isentropic vortex described in the appendix A.1 and we consider the first time step using the AUSMDV flux function and a first-order discretization in space. In Figure 5.13 (left), we see that the BiCGSTAB variants converge the fastest, but run into problems at 10^{-10}. GMRES(25) and GMRES have the same convergence behavior, thus restarting has no detrimental effect. However, the Jacobian-free variant JF-GMRES(25) starts having problems after the second restart.

Now, this test is fair in that all methods solve the same system. However, the more realistic situation is a second-order discretization in space. Therefore, we repeat the experiment with a linear reconstruction using the Barth-Jesperson limiter. Now, the schemes with a matrix solve a slightly different system with the same matrix as in the first test, but a different right-hand side, corresponding to a method of Newton-type. The Jacobian-free schemes approximately solve the system corresponding to the second-order Jacobian. As can be seen in Figure 5.13 (right), there is very little difference in convergence behavior for the schemes with matrix to the first test case. This is expected, since the system matrix is the same. On the other hand, the Jacobian-free schemes converge slower, which can be attributed to the more difficult system. Now, JF-BiCGSTAB performs worse than JF-GMRES.

We now repeat these experiments with CFL 1 and CFL 20 (see Figure 5.14) and additionally for the wind turbine problem (described in the appendix A.3) with CFL 5 and 20 (see Figure 5.15). It becomes apparent that the larger

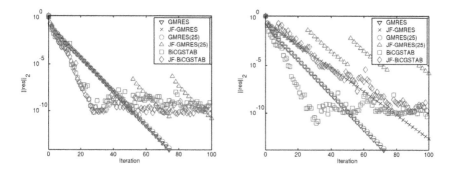

FIGURE 5.13: Norm of the relative residual over iterations for different Krylov subspace schemes for the Shu-Vortex problem at CFL=5 on the basis of a first-order FV scheme (left) and a second-order FV scheme (right).

the CFL number, the more difficult it is for the Krylov subspace methods to solve the system. In fact, Krylov subspace method, in their base form become quickly unusable for reasonable CFL numbers. This points out the necessity of using preconditioning, which is explained in the next chapter. This corresponds the observations about the system matrix made so far. Actually, for large CFL numbers, JF-BiCGSTAB is not a good method and may even diverge. This is in line with the results of Meister, that for the case with a Jacobian, BiCGSTAB performs best [196] and those of McHugh and Knoll where CGS and BiCGSTAB do not perform well in the JFNK setting [193]. Thus, we do not consider BiCGSTAB in the next experiments.

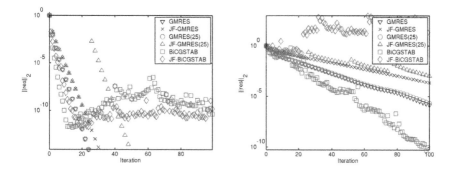

FIGURE 5.14: Comparison of different Krylov subspace schemes for the Shu Vortex problem at CFL=1 (left) and CFL=20 (right).

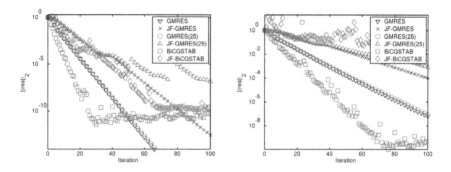

FIGURE 5.15: Comparison of different Krylov subspace schemes for the wind turbine problem at CFL=5 (left) and CFL=20 (right).

5.11   Comparison of variants of Newton's method

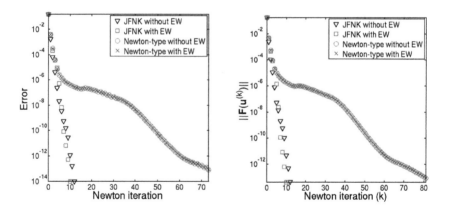

FIGURE 5.16: Illustration of Newton's method. Relative errors (left) and relative residuals (right).

To compare the different variants of Newton's method, we consider the first nonlinear system appearing when solving the Shu vortex problem A.1 using the implicit Euler method with $CFL = 0.7$. The linear systems are solved using GMRES up to machine accuracy. In Figure 5.16, the convergence curves for different variants of Newton's method are shown. In the left picture, the relative errors are shown, which have been obtained by solving the nonlinear system up to machine accuracy, storing the solution and repeating the

Newton loop. The figure on the right shows the relative residuals. As can be seen, the method with a first-order Jacobian is first-order convergent, as expected. If the Eisenstat-Walker strategy is used, this hardly changes errors or residuals and thus leads to the same order of convergence. The JFNK method exhibits second-order convergence and the Eisenstat-Walker strategy is able to obtain that as well, where again, errors and residuals are hardly changed. This demonstrates that the strategy used is indeed very reasonable.

Furthermore, for this problem, the JFNK scheme obtains quadratic convergence up to machine accuracy, which means that the constant c from theorem 10 is very small. The other thing that can be seen is that for this case, the residual is an excellent indicator of the error, being larger by less than a factor of 5.

Now, we estimate the radius of convergence by using the canonical initial guess $\mathbf{u}^{(0)} = \mathbf{u}_n$ and determine the minimal time step Δt at which the Newton iteration diverges. Intuitively, one assumes that when we increase Δt, $\mathbf{u}_{n+1}$ gets further away from $\mathbf{u}_n$. However, this need not be the case. For example for a periodic flow, it might be that an increase in Δt causes $\mathbf{u}_{n+1}$ to get closer to $\mathbf{u}_n$. Therefore, we choose as test case the Shu vortex where we know that the solution essentially moves to the right with time. The results can be seen in Table 5.6. As can be seen, the methods of Newton type with large errors

JFNK-FT	JFNK-EW	Newton-type-FT	Newton-type-EW
3.0	4.1	0.7	0.9

TABLE 5.6: Upper bounds of convergence radius for Shu vortex problem in terms of CFL numbers. FT stands for a fixed tolerance, EW for Eisenstat-Walker.

in the Jacobian lead to a smaller radius of convergence, whereas the JFNK methods have a radius of convergence that is about four times larger. The use of the Eisenstat-Walker strategy as opposed to a fixed tolerance also leads to an increase in the convergence radius.

We now compare the efficiency of the different schemes. To this end, we solve the wind turbine problem A.3 on a time interval of $10\,s$ using time-adaptive SDIRK2 with a relative and absolute tolerance of 10^{-2}. The nonlinear systems are solved with up to 40 Newton steps and the linear systems are solved using GMRES. Regarding the other tolerances, the Newton tolerance is set to $10^{-2}/5$, whereas the linear tolerance is set to $10^{-2}/50$ for the JFNK case and to 10^{-2} for the Newton type case. This is because in the latter, we only expect first-order convergence of the Newton method and thus, it is not necessary to solve the linear systems to such a high accuracy.

In Table 5.7, the total number of GMRES iterations and the total CPU time are shown. The computations are performed on one CPU of an Opteron Quad twelve-core 6168 machine with 1.9 GHz. The computation for the method of Newton-type with a fixed tolerance was stopped after more than

	JFNK-FT	JFNK-EW	Newton-type-FT	Newton-type-EW
Iter.	379,392	348,926	-	1,121,545
CPU in s	64,069	61,337	-	348,174

TABLE 5.7: Comparison of efficiency of different Newton variants.

50,000 time steps needing more than 4 million GMRES iterations and still not having computed more than 0.6 s of real time. With the Eisenstat-Walker strategy, the computations take more than five times longer than with the JFNK methods. So in both cases, the use of the Eisenstat-Walker strategy leads to a speedup, although it is less pronounced for the JFNK methods. Thus, the JFNK method is faster than the method of Newton-type, which means that the fastest method is the JFNK method with the Eisenstat-Walker strategy. This can be attributed to a number of factors playing together.

First, as just demonstrated, the JFNK method has a larger convergence radius than the method of Newton type. This means that the control that causes the time step to be repeated with a smaller time step when Newton's method fails the tolerance test after the maximal number of iterations, kicks in less often, and actually leads to larger possible time steps. Second, the JFNK method is second-order convergent, needing less Newton steps, thus being more efficient and having again the same effect as the first issue. Third, the Eisenstat-Walker strategy reduces the tolerance at which the linear systems are solved. This is good in itself, since it leads to less Krylov subspace iterations, but there is an added benefit when using GMRES: Since GMRES needs more storage and more computational effort with every iteration, it is extremely fast for small tolerances. Finally, we have seen in the last section that GMRES in the JFNK context works better if we avoid restarting. This is achieved if the tolerances are such that GMRES terminates before the maximal dimension of the Krylov subspace is reached.

Chapter 6

Preconditioning linear systems

As discussed in the last chapter, the speed of convergence of Krylov subspace methods depends strongly on the matrix and matrices arising from the discretization of PDEs are typically such that Krylov subspace methods converge slowly. Therefore, an idea called preconditioning is used to transform the linear equation system into an equivalent one to speed up convergence:

$$\mathbf{P}_L^{-1}\mathbf{A}\mathbf{P}_R^{-1}\mathbf{x}^P = \mathbf{P}_L^{-1}\mathbf{b}, \quad \mathbf{x} = \mathbf{P}_R^{-1}\mathbf{x}^P.$$

Here, $\mathbf{P}_L^{-1}$ and $\mathbf{P}_R^{-1}$ are invertible matrices, called a left and right preconditioner, respectively, that approximate the system matrix in a cheap way. The optimal preconditioner in terms of accuracy is $\mathbf{A}^{-1}$, since then we would just need to solve a system with the identity matrix. However, that's not efficient. In terms of cost, the optimal preconditioner is the identity, which comes for free, but also has no effect on the number of iterations. This idea has turned out to be extremely powerful and allows preconditioned Krylov subspace methods to be the methods of choice for large classes of discretized PDEs. In fact, the choice of preconditioner is more important than the choice of Krylov subspace method.

Often, the preconditioner is not given directly, but implicitly by its inverse. Then, the application of the preconditioner corresponds to the solution of a linear equation system. If chosen well, the speedup of the Krylov subspace method is enormous and therefore, the choice of the preconditioner is more important than the specific Krylov subspace method used.

Preconditioning can be done very easily in Krylov subspace methods. Every time a matrix vector product $\mathbf{A}\mathbf{v_j}$ appears in the original algorithm, the right preconditioned method is obtained by applying the preconditioner $\mathbf{P}_R^{-1}$ to the vector $\mathbf{v}_j$ in advance and then computing the matrix vector product with $\mathbf{A}$. For left preconditioning the preconditioner is applied afterwards instead. Hence, left preconditioned GMRES works in the Krylov subspace

$$\mathcal{K}_k(\mathbf{P}^{-1}\mathbf{A}, \mathbf{r}_0^P) = \text{span}\{\mathbf{r}_0^P, \mathbf{P}^{-1}\mathbf{A}\mathbf{r}_0^P, ..., (\mathbf{P}^{-1}\mathbf{A})^{k-1}\mathbf{r}_0^P\},$$

whereas the Krylov subspace generated by right preconditioning is

$$\mathcal{K}_k(\mathbf{A}\mathbf{P}^{-1}, \mathbf{r}_0) = \text{span}\{\mathbf{r}_0, \mathbf{A}\mathbf{P}^{-1}\mathbf{r}_0, ..., (\mathbf{A}\mathbf{P}^{-1})^{k-1}\mathbf{r}_0\}.$$

Note that right preconditioning does not change the initial residual, because

$$\mathbf{r}_0 = \mathbf{b}_0 - \mathbf{A}\mathbf{x}_0 = \mathbf{b}_0 - \mathbf{A}\mathbf{P}_R^{-1}\mathbf{x}_0^P,$$

therefore the computation of the initial residual can be done without the right preconditioner. However, when the tolerance criterion is fulfilled, the right preconditioner has to be applied one last time to change back from the preconditioned approximation to the unpreconditioned. On the other hand, a left preconditioner has to be applied once initially, but not afterwards. This means that left preconditioning changes the residual.

As mentioned in section 5.8.1, there is no satisfying convergence theory for nonnormal matrices as we have here. Therefore, the preconditioner has to be chosen by numerical experiments and heuristics. An overview of preconditioners with special emphasis on application in flow problems can be found in the book of Meister [197] and the study in the context of compressible flow by Meister and Vömel [199].

6.1   Preconditioning for JFNK schemes

Regarding preconditioned versions of the Jacobian-free matrix vector product approximation (5.60), we obtain for left preconditioning

$$\mathbf{P}^{-1}\mathbf{A}\mathbf{q} \approx \mathbf{P}^{-1}\left(\frac{\mathbf{F}(\bar{\mathbf{u}} + \epsilon\mathbf{q}) - \mathbf{F}(\bar{\mathbf{u}})}{\epsilon}\right) \tag{6.1}$$

whereas for right preconditioning, we have

$$\mathbf{A}\mathbf{P}^{-1}\mathbf{q} \approx \frac{\mathbf{F}(\bar{\mathbf{u}} + \epsilon\mathbf{P}^{-1}\mathbf{q}) - \mathbf{F}(\bar{\mathbf{u}})}{\epsilon}. \tag{6.2}$$

Thus, preconditioning can be implemented in exactly the same way as for the case with Jacobian. However, the construction of the preconditioner becomes a problem, since we do not have a Jacobian anymore. In fact, while the JFNK scheme as presented so far can be implemented into an existing code with little extra work and in particular, no new data structures, it is here that things can become complicated. Regarding this, there are two approaches:

1. Compute the Jacobian nevertheless, then compute the preconditioner and store that in place of the Jacobian.

2. Use preconditioners that need only parts of or no part of the Jacobian, thus leading to extremely low storage methods.

The first strategy still reduces the storage needed by one matrix compared to the case with Jacobian. It is for example necessary when using ILU preconditioning, which will be explained in the next section. The second strategy severely limits the set of possible preconditioners. Jacobi would be one example, since this just uses the diagonal. Another example would be multigrid methods with appropriate smoothers, in particular Jacobi or RK smoothing.

6.2 Specific preconditioners

6.2.1 Block preconditioners

As discussed in section 5.2, the matrix we are concerned with is a block matrix and if this structure is respected by the preconditioner, the convergence speed is increased. This is done by interpreting all methods if possible as block methods with an appropriately chosen block size. The main difference of preconditioners for finite volume methods as opposed to DG methods is the size of the blocks, which is significantly larger for DG schemes. This makes an efficient treatment of the blocks imperative for a successful implicit DG scheme, whereas it is only beneficial for a finite volume scheme.

In the case of the DG-SEM method, there are different block sizes that can be used (compare Figure 5.1). First of all, there are the small blocks corresponding to one degree of freedom which are of the same size as the number of equations, namely $d + 2$. Then, this can be increased to include all degrees of freedom in x_1-direction, leading to a block of size $(d+2)(p+1)$. Of course, the x_2 direction can be included as well, until if also the x_3 direction is included, the block corresponds to all the unknowns in one cell. For the modal-nodal DG scheme, we later suggest the ROBO-SGS method, which exploits the specific block structure of the hierarchical basis.

6.2.2 Stationary linear methods

The iterative methods explained in section 5.5.1, although inferior to Krylov subspace methods for the solution of linear systems, turn out to provide useful preconditioners. They were based on the splitting

$$\mathbf{A} = (\mathbf{A} - \mathbf{N}) + \mathbf{N},$$

giving rise to the fixed point method (5.12):

$$\mathbf{x}^{(k+1)} = (\mathbf{I} - \mathbf{N}^{-1}\mathbf{A})\mathbf{x}^{(k)} + \mathbf{N}^{-1}\mathbf{b}.$$

Thus, they are based on approximating $\mathbf{A}^{-1}$ via the matrix $\mathbf{N}^{-1}$. Therefore, $\mathbf{N}^{-1}$ can be used as a preconditioner. In particular the symmetric block Gauss-Seidel (SGS) method is a very good preconditioner for compressible flow problems. One application of SGS as a preconditioner to the vector $\mathbf{q}$ corresponds to solving the equation system (see (5.16))

$$(\mathbf{D} + \mathbf{L})\mathbf{D}^{-1}(\mathbf{D} + \mathbf{U})\mathbf{x} = \mathbf{q}. \qquad (6.3)$$

For the purpose of preconditioning, one iteration is completely sufficient. The blocks of $\mathbf{L}$ and $\mathbf{U}$ can be computed when required one at a time, so it is not necessary to store the complete matrix. Only the diagonal blocks which appear several times are computed in advance and stored.

Another simple choice for a splitting method would be Jacobi-preconditioning, corresponding to $\mathbf{P}^{-1} = \mathbf{D}^{-1}$ (see 5.18), where we need only the diagonal blocks, thus reducing the cost of applying the preconditioner and a preconditioner that is simple to implement. For finite-volume methods, the blocks are of the size of the number of unknowns on the PDE. These blocks are too small to contain enough information about the matrix, leading to a bad preconditioner. For DG methods, the choice can be made to define a diagonal block to correspond to all unknowns in one cell. This makes the Jacobi preconditioner much more expensive to apply, but also more powerful.

A common technique to improve the efficiency of a splitting-related preconditioner is relaxation, which corresponds for SGS to:

$$\frac{1}{\omega(2 - \omega)}(\mathbf{D} + \omega\mathbf{L})\mathbf{D}^{-1}(\mathbf{D} + \omega\mathbf{U})\mathbf{x} = \mathbf{q}.$$

Unfortunately, it is difficult to find a choice of ω that leads to consistently better results and significant speedup can not be expected.

A linear multigrid method gives rise to a preconditioner as well, given by the recursion

$$\mathbf{N}_{MG,0}^{-1} = \mathbf{N}_{S,0}^{-1} \qquad (6.4)$$

$$\mathbf{N}_{MG,l}^{-1} = \mathbf{M}_{S,l}(\mathbf{N}_{S,l}^{-1} - P_{l,l-1}\mathbf{N}_{MG,l-1}^{-1}\mathbf{R}_{l-1,l}\mathbf{A}_l\mathbf{N}_{S,l}^{-1} + P_{l-1,l}\mathbf{N}_{MG,l-1}^{-1}\mathbf{R}_{l-1,l})$$
$$+ \mathbf{N}^{-1}, \, l = 1, ..., l_{\max}. \qquad (6.5)$$

6.2.3   ROBO-SGS

In [34], the low-memory SGS preconditioner with a reduced offblock order (ROBO-SGS) is suggested for use with modal DG schemes (see section 3.7.1). This exploits that the basis is hierarchical in that for the offdiagonal blocks $\mathbf{L}$ and $\mathbf{U}$ in (6.3), only the parts corresponding to a lower order discretization are computed (see Figure 6.1). The number of entries of the offdiagonal blocks in the preconditioner is then $N \cdot (d + 2) \times \hat{N} \cdot (d + 2)$, where $\hat{N}$ is the dimension of the lower order polynomial space of order $\hat{p}$ chosen for the offdiagonal blocks. We denote the corresponding preconditioner as ROBO-SGS-$\hat{p}$; for example, ROBO-SGS-1 is the SGS preconditioner where in the offdiagonal blocks, only the terms of order 1 and 0 are considered. Thus, the unknowns whose contributions are incorporated into the preconditioner have a physical meaning even for low orders.

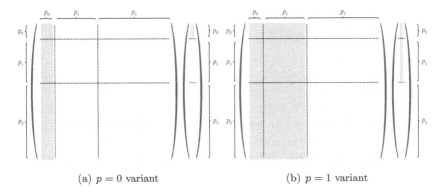

(a) $p = 0$ variant   (b) $p = 1$ variant

FIGURE 6.1: Reduced versions of the off-block Jacobians, $p = 0$ and $p = 1$ variants. (Credit: M. Ferch.)

While this results in a decreased accuracy of the preconditioner, the memory requirements and the computational cost of the application become better, the less degrees of freedom of the neighboring cells one takes into account. We consider the significant savings in memory to be even more important for 3D simulations since there memory is usually the main limitation for high-order DG schemes.

6.2.4 ILU preconditioning

Another important class of preconditioners are block incomplete LU (ILU) decompositions, where the blocks correspond to the small units the Jacobian consists of. A thorough description can be found in the book of Saad [241]. The computation of a complete LU decomposition is quite expensive and in general leads to a dense decomposition, also for sparse matrices. By prescribing a sparsity pattern, incomplete LU decompositions can be defined, which consist of an approximate factorization $\mathbf{LU} \approx \mathbf{A}$. The larger the sparsity pattern, the better the approximation and the more powerful the preconditioner, but the more expensive are application and computation. The application of such a decomposition as a preconditioner is done by solving the corresponding linear equation system using forward-backward substitution.

Given a sparsity pattern $M \subset \{(i,j)|i \neq j, 1 \leq i, j \leq m\}$, the algorithm can be written as

- For $k = 1, ..., m - 1$

-      for $i = k + 1, ..., m$ and if $(i, k) \notin M$

-         $a_{ik} = a_{ik}/a_{kk}$

-         for $j = k + 1, ..., m$ and if $(i, j) \notin M$

- $a_{ij} = a_{ij} - a_{ik}a_{kj}$

- end for

- end for

- end for

The question remains how the sparsity pattern should be chosen. A widely used strategy is the level of fill l, leading to the preconditioner ILU(l). The level of fill is a recursively defined measure for how fill-in much beyond the original sparsity pattern is allowed: Using the sparsity pattern of **A** corresponds to level 0. The sparsity pattern for level 1 is obtained by computing ILU(0), multiplying the corresponding **L** and **U** matrices and using the nonzero entries of that product. Level 2 is obtained by computing ILU(1), multiplying the corresponding matrices and so on. For the purpose of memory allocation, these patterns can be determined in advance based on the sparsity pattern of A, meaning that the intermediate matrices and multiplications do not need to be generated. Those decompositions with higher levels of fill are very good black box preconditioners for flow problems. However, they also have large memory requirements. Thus remains ILU(0), which has no additional level of fill beyond the sparsity pattern of the original matrix **A**.

Another widely used method is ILUT, where the T stands for drop tolerance. There, the sparsity pattern is determined on the fly, by dropping an entry in the **LU** factorization, if it is below a drop tolerance ϵ. This leads to a preconditioner with a good compromise between accuracy and efficiency. However, the implementation is much more difficult, due to the variable sparsity pattern.

Note that as GS and SGS, the ILU depends on the ordering of the unknowns, as does its performance. This is because the ordering of unknowns changes the amount of fill-in of the exact LU decomposition and thus also the errors introduced by the ILU. Strategies to do this are for example reverse Cuthill-McKee ordering [241], as well as the physical reordering suggested for SGS in [199]. There, unknowns are renumbered along planes normal to the flow direction, thus the numeration partly reflects the physical flow of information.

6.2.5   Multilevel preconditioners

Another possibility is to use linear multilevel schemes as preconditioners. Using nonlinear schemes as preconditioners is covered in the next subsection. First of all, the linear multigrid methods as described in section 5.6 can be applied. Then, the multi-p-method can be applied to the linear problem. Finally, there are other generalizations of multigrid methods to multilevel ideas, for example algebraic multigrid. In the finite volume context, this has been analyzed in [191] and found to be competitive when compared to FAS multigrid or using Newton with Multigrid as a linear solver. A number of approaches have been tried in the context of DG methods, e.g., multigrid methods and multi-p methods with different smoothers (see [210, 192] for steady Euler flows, [35] for unsteady Euler flows), as well as a variant by Persson and Peraire [224], where a two-level multi-p-method is used with ILU(0) as a postsmoother.

The 2-grid preconditioner is given by (compare (5.43))

$$\mathbf{P}_{2G_l}^{-1} = \mathbf{N}_{S_l}^{-1} + \mathbf{P}_{l,l-1}\mathbf{A}_{l-1}^{-1}\mathbf{R}_{l-1,l}(\mathbf{I} - \mathbf{A}_l\mathbf{N}_{S_l}^{-1}).$$

In the multilevel version, $\mathbf{A}_{l-1}^{-1}$ is replaced by a recursive application of the preconditioner. To implement this in a Jacobian-free context, the matrix vector product is replaced by the finite difference approximation. This also works on the coarse grids, if the operator there is defined by the coarse grid discretization, instead of a transformed fine grid matrix, which is the standard for agglomeration multigrid. The prolongation and restriction are sparse matrices, and can be implemented in a matrix-free fashion, which means that the coefficients are hard coded.

This leaves the smoother as the crucial component. If this can be implemented in a Jacobian-free way, then the whole multigrid preconditioner is Jacobian-free. For DG methods, good options are block Jacobi smoothers, when the blocks are chosen cellwise. These are rather costly, though. Later, we will test the preconditioner of Persson in Peraire with the difference, that we use an ILU presmoother instead of postsmoothing. We call this method ILU-CSC (for coarse scale correction).

6.2.6 Nonlinear preconditioners

If there is an already existing nonlinear multigrid code with dual time stepping for unsteady flows, it seems natural to use that method as a preconditioner in a JFNK method (5.60), since this gives a low storage preconditioner at no additional implementation cost. This approach was first tried for steady problems by Wigton, Yu, and Young in 1985 [303], later by Mavriplis [191] and then for unsteady problems by Bijl and Carpenter [22]. In [38] and [37], Birken and Jameson analyze the use of general nonlinear solvers as preconditioners. This section follows these two articles. It turns out that the use of nonlinear left preconditioning alters the equation system in a nonequivalent way, leading to a stall in Newton convergence. For an overview on more general methods, see [50].

Alternatively, we can use nonlinear right preconditioning:

$$\mathbf{A}\mathbf{P}^{-1}\mathbf{q} \approx \frac{\mathbf{F}(\underline{\mathbf{u}}^{(k)} + \epsilon\mathbf{P}^{-1}\mathbf{q}) - \mathbf{F}(\underline{\mathbf{u}}^{(k)})}{\epsilon},$$

where $\mathbf{P}$ is now a nonlinear operator. The following problems occur:

1. GMRES uses basisvectors of the solution space, thus these vectors correspond to $\Delta\underline{\mathbf{u}}$. However, the FAS multigrid does not work on differences of solutions, but on the solution vector $\underline{\mathbf{u}}$; therefore, it cannot be used.

2. Since $\mathbf{P}^{-1}$ might be variable, we do not really know what the proper backtransformation would be to transform the preconditioned solution back to the unpreconditioned space.

The second problem is solved by the flexible GMRES method (FGMRES) of Saad [240], which allows to use a different preconditioner in every iteration. However, since FMGRES is just a slight modification of GMRES, it still has the first problem. Both problems are solved by GMRES-* [285] of van der Vorst and Vuik, which allows nonlinear right preconditioning. There, the * represents the right preconditioner. Thus, we could have GMRES-SGS or GMRES-DTS (for Dual Time Stepping). GMRES-* is mathematically equivalent to GCR, which is in the fixed preconditioner case mathematically equivalent to GMRES. The preconditioner in GMRES-* is applied by replacing the line $\mathbf{p}_k = \mathbf{r}_k$ in the GCR algorithm with the application of the preconditioner to the residual $\mathbf{r}_k$ and the storing of the result in the search direction $\mathbf{p}_k$. Thus, the preconditioner works with residual vectors and nonlinear right preconditioning is applied by

$$\mathbf{P}^{-1}\mathbf{r}_m \approx \mathbf{P}^{-1}\mathbf{F}(\underline{\mathbf{u}}^{(k)} + \mathbf{x}_m) = \underline{\mathbf{u}}^{(k)} + \mathbf{x}_m - \mathbf{N}(\underline{\mathbf{u}}^{(k)} + \mathbf{x}_m). \qquad (6.6)$$

Here, $\mathbf{N}(\underline{\mathbf{u}})$ is some nonlinear method for the solution of the original nonlinear equation $\mathbf{F}(\underline{\mathbf{u}}^{n+1}) = \mathbf{0}$, for example FAS multigrid. The definition is motivated by remembering that

$$\mathbf{r}_m = \mathbf{A}\mathbf{x}_m - \mathbf{b} = \mathbf{F}(\underline{\mathbf{u}}^{(k)}) + \frac{\partial \mathbf{F}}{\partial \underline{\mathbf{u}}}\bigg|_{\underline{\mathbf{u}}^{(k)}} (\underline{\mathbf{u}}^{(k+1)} - \underline{\mathbf{u}}^{(k)}) \approx \mathbf{F}(\underline{\mathbf{u}}^{(k)} + \mathbf{x}_m).$$

This is a truly nonlinear method, which does not have the same problem as the left preconditioner of changing the right-hand side of the Newton scheme.

The residual history of left and right preconditioning for different test cases is shown in Figure 6.2. The same test cases as above were used, where the system solved is that from the first Newton iteration. As can be seen, the left preconditioner improves the convergence speed in the unsteady case, but not in the steady case. There, $\mathbf{N}(\underline{\mathbf{u}})$ is close to the identity. The right preconditioner on the other hand, which does not lead to a stall in the Newton iteration, also does not improve the convergence speed much. In the case of steady flow, it even leads to a loss of speed of convergence. We conclude that this type of nonlinear preconditioning is not advisable.

6.2.7   Other preconditioners

There is a multitude of other preconditioners that have been suggested over time. One example are polynomial preconditioners. There, it is used that if $\rho(\mathbf{I} - \mathbf{A}) < 1$, the inverse of $\mathbf{A}$ can be represented by its Neumann series

$$\mathbf{A}^{-1} = \sum_{k=0}^{\infty} (\mathbf{I} - \mathbf{A})^k.$$

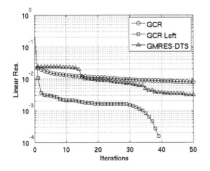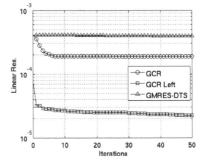

FIGURE 6.2: Linear Residual versus Iteration number for one linear system of unsteady viscous flow (left) and steady inviscid flow (right). (Credit: First published in [37], published by the American Mathematical Society.)

Thus,

$$\mathbf{P}^{-1} = \sum_{k=0}^{m} (\mathbf{I} - \mathbf{A})^k \qquad (6.7)$$

or more generally

$$\mathbf{P}^{-1} = \sum_{k=0}^{m} \Theta_k (\mathbf{I} - \mathbf{A})^k \qquad (6.8)$$

with $\Theta_k \in \mathbb{R}$ can be used as a preconditioner.

Furthermore, there are physical preconditioners that use the discretization of simpler problems that still incorporate core issues of stiffness like low Mach numbers, diffusion or grid-induced stiffness to define a preconditioner [219]. Here, it is important to include the sources of stiffness, as for example using the Euler equations to define a preconditioner does not lead to a fast method. This is because in this case, the simplified model is less stiff than the approximated model, the Navier-Stokes equations.

Other ideas would be the use of the alternate direction implicit (ADI) splitting for structured grids [6] or the Sparse approximate inverse (SPAI) preconditioner aimed at parallel computations [115].

6.2.8 Comparison of preconditioners

We now compare the effect of different preconditioners on the convergence behavior of Krylov subspace methods for a single system. Since this does not take into account the performance of the overall solution scheme for an unsteady flow problem, we cannot say what the best preconditioner is. Appropriate results will be discussed in the next chapter.

We first analyze the influence of ordering on convergence. To this end, we consider the flow around a cylinder at Mach 10 from appendix A.2. We solve this using the finite volume code TAU_2D (see chapter 7). We solve the first system when using the implicit Euler method with $CFL = 2.0$ with ILU preconditioned Jacobian-free GMRES and GMRES. As can be seen in Figure 6.3, the physical reordering section 6.2.4 leads to a small, but noticeable improvement in convergence behavior for the case with a Jacobian, but doesn't change much for the Jacobian-free case, where we have to remember that the linear system solved in that case is the second-order system, being more difficult. For other cases, with smaller Mach numbers, the physical renumbering strategy has no effect.

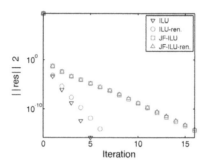

FIGURE 6.3: Effect of physical renumbering.

To evaluate the effect of using different preconditioners for finite volume schemes, we consider the wind turbine problem A.3 and solve it using the in-house FV code TAU_2D. We solve the first system when using the implicit Euler method. In Figure 6.4, we can see that Jacobi preconditioning has essentially no effect on the convergence speed, in fact it is barely able to lead to a speedup at all. By contrast, ILU preconditioning increases convergence speed by more than a factor of 2. As before, for the more difficult system with $CFL = 20$, even the ILU preconditioned scheme needs a significant number of iterations. However, note that for engineering accuracies of 10^{-3} or 10^{-5}, ILU is able to prevent a restart.

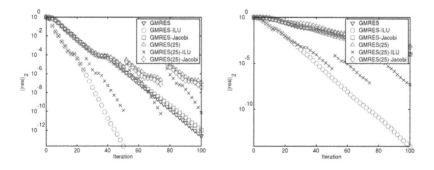

FIGURE 6.4: Comparison of different preconditioners for FV. CFL=5 (left) and CFL=20 (right).

For the case of DG methods in parallel, we compare Jacobi, ILU, ILU-CSC and ROBO-SGS-0 till ROBO-SGS-4 for a fifth-order polymorphic modal-nodal discretization of the sphere vortex shedding problem A.4. We use eight CPUs in a domain decomposition fashion, as explained in section 7.3. The system solved is the first system arising in the Newton scheme for the first nonlinear system during the first time step of an ESDIRK4 time integration scheme with a time step size of $\Delta t = 0.065$. As a solver, Jacobian-Free GM-RES(20) is employed. The relative residuals are shown in Figure 6.5.

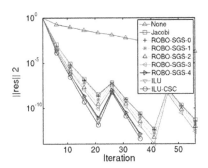

FIGURE 6.5: Comparison of different preconditioners for DG discretization.

As can be seen, the most powerful preconditioner is ILU-CSC, followed by second most powerful preconditioner ILU. Then, we have ROBO-SGS-4, which is almost as powerful as ILU, and then the other ROBO-SGS preconditioner in decreasing order. As expected, Jacobi is the least powerful preconditioner. Regarding efficiency, the most powerful preconditioner is not necessarily the best one. This is discussed in section 7.5.

6.3 Preconditioning in parallel

When doing computations in parallel, a number of things about preconditioners have to be rethought. In particular, GS, SGS, and ILU are inherently sequential methods where some parts of the computations can be carried out only after some other parts. This is not the case for Jacobi and it is also not the case for multigrid methods, provided the smoothers are appropriately chosen. For example, RK smoothers work well in parallel, whereas SGS obviously does not. For DG methods, because of the large block sizes, block Jacobi is a good default choice of preconditioner or smoother that is trivial to parallelize.

An important idea to mention here is a domain decomposition approach for the preconditioner [227] and [278]. This means that the spatial domain is decomposed into several connected subdomains, each of which is assigned to one processor. The preconditioner is then applied locally. When using this approach, care has to be taken since the performance of SGS and ILU in particular depend on the choice of subdomains. Additionally, the question is how much communication one allows over the domain boundaries. One example of a domain decomposition method is the Dirichlet-Neumann iteration explained chapter 7.

6.4   Sequences of linear systems

In the context of unsteady flows, we always have a sequence of linear systems with slowly varying matrices and right-hand sides. To obtain an efficient scheme, it is mandatory to exploit this structure. One example for this was the idea of the simplified Newton method, where the Jacobian is not recomputed in every time step. Since setup costs of preconditioners are often high, it is necessary to abstain from computing the preconditioner for every matrix in the sequence. Instead, one of the following strategies should be used.

6.4.1   Freezing and recomputing

The most common strategy is to compute the preconditioner for the first system and then use it for a number of the following system (freezing), for example by defining a recomputing period l, where after l time steps, the preconditioner is constructed again. Often, the preconditioner is frozen completely either from the beginning or after a number of time steps [199]. For ILU and Jacobi preconditioning, this is a natural thing, but for SGS, we have another choice. There, it is possible to store only the diagonal and compute the off diagonal blocks on the fly, implying that they have to be computed anew for every system. Alternatively, if storage is not a problem, the off diagonal blocks can be stored, allowing to use freezing and recomputing as a strategy.

6.4.2   Triangular preconditioner updates

An idea that aims at reusing information from previous linear systems are preconditioner updates. The technique we base our updates on was suggested originally by Duintjer Tebbens and Tuma in [78]. It was then reformulated and refined for block matrices [32, 31] and later put into the JFNK context [79]. This strategy has the advantage that it works for any matrix, is easy to implement, parameter-free, and with only a small overhead. However, it requires storing not only the matrix and an ILU preconditioner, but also a reference matrix with a reference ILU preconditioner, which means that the storage requirements are high. Nevertheless, it is a powerful strategy if storage is not an issue and has been applied also outside of CFD with good results [297]. This section follows [32].

In addition to a system $\mathbf{A}\mathbf{x} = \mathbf{b}$ with a block ILU(0) (BILU(0)) preconditioner $\mathbf{P} = \mathbf{LDU} = \mathbf{LU_D}$, let $\mathbf{A}^+\mathbf{x}^+ = \mathbf{b}^+$ be a system of the same dimension with the same sparsity pattern arising later in the sequence and denote the difference matrix $\mathbf{A} - \mathbf{A}^+$ by $\mathbf{B}$. We search for an updated block ILU(0) preconditioner $\mathbf{P}^+$ for $\mathbf{A}^+\mathbf{x}^+ = \mathbf{b}^+$. Note that this implies that all matrices have the same sparsity pattern and thus, after a spatial adaptation, the scheme has to be started anew.

We have

$$\|\mathbf{A} - \mathbf{P}\| = \|\mathbf{A}^+ - (\mathbf{P} - \mathbf{B})\|,$$

hence the level of accuracy of $\mathbf{P}^+ \equiv \mathbf{P} - \mathbf{B}$ for $\mathbf{A}^+$ is the same, in the chosen norm, as that of $\mathbf{P}$ for $\mathbf{A}$. The updating techniques from [78] are based on the *ideal* updated preconditioner $\mathbf{P}^+ = \mathbf{P} - \mathbf{B}$. If we would use it as a preconditioner, we would need to solve systems with $\mathbf{P} - \mathbf{B}$ as system matrix in every iteration of the linear solver. For general difference matrices $\mathbf{B}$, these systems would be too hard to solve. Therefore, we consider cheap approximations of $\mathbf{P} - \mathbf{B}$ instead.

Under the assumption that $\mathbf{P}-\mathbf{B}$ is nonsingular, we approximate its inverse by a product of triangular factors which are easier to invert. In particular, we will end up with using either only the lower triangular or the upper triangular part of $\mathbf{B}$. First, we approximate $\mathbf{P} - \mathbf{B}$ as

$$\mathbf{P} - \mathbf{B} = \mathbf{L}(\mathbf{U_D} - \mathbf{L}^{-1}\mathbf{B}) \approx \mathbf{L}(\mathbf{U_D} - \mathbf{B}), \tag{6.9}$$

or by

$$\mathbf{P} - \mathbf{B} = (\mathbf{LD} - \mathbf{B}\mathbf{U}^{-1})\mathbf{U} \approx (\mathbf{LD} - \mathbf{B})\mathbf{U}. \tag{6.10}$$

Next, we replace $\mathbf{U_D} - \mathbf{B}$ or $\mathbf{LD} - \mathbf{B}$ by a nonsingular and easily invertible approximation. Following [32], we use

$$\mathbf{U_D} - \mathbf{B} \approx btriu(\mathbf{U_D} - \mathbf{B}),$$

or

$$\mathbf{LD} - \mathbf{B} \approx btril(\mathbf{LD} - \mathbf{B}),$$

where $btriu$ and $btril$ denote the block upper and block lower triangular parts (including the main diagonal), respectively. Putting the two approximation steps together, we obtain two possible updated preconditioners in the form

$$\mathbf{P}^+ = \mathbf{L}(\mathbf{U_D} - btriu(\mathbf{B})) \tag{6.11}$$

and

$$\mathbf{P}^+ = (\mathbf{LD} - btril(\mathbf{B}))\mathbf{U}. \tag{6.12}$$

These can be obtained very cheaply. They ask only for subtracting block triangular parts of $\mathbf{A}$ and $\mathbf{A}^+$ (and for saving the corresponding block triangular part of $\mathbf{A}$). In addition, as the sparsity patterns of the factors from the BILU(0) factorization and from the block triangular parts of $\mathbf{A}$ (and $\mathbf{A}^+$) are identical, both backward and forward substitution with the updated preconditioners are as cheap as with the frozen preconditioner $\mathbf{LU_D} = \mathbf{LDU}$.

Regarding the distance of the updated preconditioners (6.11) and (6.12) to the ideal preconditioner, we can deduce from the two approximations we make, that it is mainly influenced by the following two properties. The first is closeness of $\mathbf{L}$ or $\mathbf{U}$ to the identity. If matrices have a strong diagonal, the diagonal dominance is in general inherited by the factors $\mathbf{L}$ and $\mathbf{U}$ [19, 18], yielding reasonable approximations of the identity. The second property is a block triangular part containing significantly more relevant information than the other part, which means $btril(\mathbf{B})$ or $btriu(\mathbf{B})$ is a useful approximation of $\mathbf{B}$.

We now describe how to apply updated preconditioners in the solution process. A first issue is the choice between (6.11) and (6.12). In [32], three different criteria are suggested and here we present the *stable update criterion*. This compares the closeness of the factors to identity, keeping the factor that is closest to the identity, which leads to more stable back and forward substitutions. This is done by comparing the norms $\|\mathbf{L} - \mathbf{I}\|$ and $\|\mathbf{U} - \mathbf{I}\|$. If the former norm is smaller, then we update the upper triangular part of the decomposition by (6.11), whereas if, on the contrary, $\mathbf{U}$ is closer to identity in some norm, we update the lower triangular part according to (6.12).

A related issue is the frequency of deciding about the update type based on the chosen criterion. On the one hand, there may be differences in the performance of (6.11) and (6.12); on the other hand, switching between the two types implies some additional costs like storage of both triangular parts of $\mathbf{B}$. Consequently, the query is used only directly after a recomputation of the BILU(0) decomposition, which takes place periodically. The chosen type of update is then used throughout the whole period. For the stable update criterion we may decide immediately which update type should be used as soon as the new BILU(0) decomposition is computed. Note that as soon as the update type is chosen, we need to store only one triangular part of the old reference matrix $\mathbf{A}$ (and two triangular factors of the reference decomposition).

As another tweak, we do not start the updating right away after a recomputation of the frozen preconditioner. This is because in the sequence of linear systems it may happen that several succeeding system matrices are very close and then the frozen preconditioner should be powerful for many subsequent systems. Denote the number of iterations of the linear solver needed to solve the first system of the period by $iter_0$. If for the $(j + 1)$st system the corresponding number of iterations $iter_j$ satisfies

$$iter_j > iter_0 + 3, \tag{6.13}$$

where the threshold 3 is chosen heuristically, we start updating.

6.4.3   Numerical results

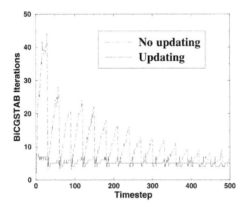

FIGURE 6.6: BiCGSTAB iterations over time-steps for a cylinder at Mach 10.

We now compare the triangular updates to freezing and recomputing for a model problem, namely the computation of the steady state around a cylinder, being hit by a Mach 10 2D Euler flow. Three thousand steps of the implicit Euler method are performed. In the beginning, a strong shock detaches from the cylinder, which then slowly moves backward through the domain until reaching the steady state position. Therefore, the linear systems are changing only very slowly during the last 2500 time steps and all important changes take place in the initial phase of 500 time steps. The initial CFL number is 5, which is increased up to 7 during the iteration. All computations were performed on a Pentium IV with 2.4 GHz. We use the finite volume code TAU_2D (see chapter 7) and BiCGSTAB to solve the linear systems. The solution is shown in Figure 6.6.

As the flow is supersonic, the characteristics point mostly in one direction and therefore, the physical reordering from section 6.2.4 leads to a situation where we obtain a matrix with a dominant triangular part. In Figure 6.6, the effect of preconditioner updates on the number of Krylov iterations is compared to a periodic recomputation strategy. As subsequent linear systems change heavily, frozen preconditioners produce rapidly increasing numbers of BiCGSTAB iterations (with decreasing peaks demonstrating the convergence to steady state). Updating, on the other hand, yields a nearly constant number of iterations per time step. The recomputing period here is 30 and the criterion used is the stable update criterion but other periods and criteria give a similar result. With freezing, 5380 BiCGSTAB iterations are performed in this part of the solution process, while the same computation with updating needs only 2611 iterations.

The performance of the updates for the whole sequence is displayed in Table 6.1. To evaluate the results, first note that the reduction of the BiCGSTAB

	Freezing/Recomp.		Stable update	
Period	Iter.	CPU in s	Iter.	CPU in s
10	10683	7020	11782	7284
20	12294	6340	12147	6163
30	13787	7119	12503	5886
40	15165	6356	12916	5866
50	16569	6709	13139	5786

TABLE 6.1: Total iterations and CPU times for supersonic flow example

iterations happens primarily in the first 500 time steps. After 500 time steps, freezing is a very efficient strategy and actually gains again on updating. As expected, the update criterion chooses to update the lower triangular part according to (6.12), because the upper triangular part is close to identity due to the numbering of the unknowns and the high Mach number. Updating is clearly better than freezing if the recomputing period is at least 20. The CPU time is decreased by about 10% in general; with the recomputing period 50, it reaches up to 20%. For high recomputing periods, the number of iterations is reduced by even more than 20%. If the ILU decomposition would have been recomputed in every step, only 11,099 BiCGSTAB iterations would be needed, but 28,583 s of CPU time.

For recomputing periods of 30 or greater, the performance of the updating strategy does not much depend on the period, which means that the solver becomes more robust in its behavior with respect to the specific choice of the updating period.

When considering different test cases, we see that the triangular updates are less powerful than for this example, if the flows are very steady. Nevertheless, the updating strategy is roughly as efficient as freezing and recomputing, even for test cases where it should perform bad, like steady low Mach number flows.

6.5   Discretization for the preconditioner

The other question, besides which specific preconditioner to use, is, which discretization we base the computation of the blocks on. In the finite volume case, the Jacobian computed is based on the first-order discretization in the first place. Now, one advantage of JFNK schemes is that the flux function can be changed easily without a need to reprogram a Jacobian. However, this is no longer the case when an ILU preconditioner is used. Therefore, one idea is to base the computation of the preconditioner on a fixed flux function, thus regaining this advantage of JFNK schemes. In [186], the van Leer flux-vector splitting is used for the preconditioner, as then the blocks can be computed quite efficiently. By contrast, the Jacobian of the AUSMDV or HLLC flux is extremely complicated and expensive to compute.

In Figure 6.7, the convergence of ILU(0) preconditioned JFNK-GMRES, where the preconditioner is based on the same flux function as used for the discretization, is compared to the case where a different discretization is used. Furthermore, the results of the scheme with Jacobian are shown. More precise, the in-house 2D-FV code TAU_2D is used with the AUSMDV flux and the computation of the Jacobian is based on AUSMDV or the van Leer flux vector splitting. The problem considered is the wind turbine test case A.3 and we show the result of the first time step of an implicit Euler calculation with $CFL = 5$. As can be seen, the method with mismatched discretizations is only slightly slower than its counterpart. This means that this strategy can be used to simplify the implementation of implicit schemes.

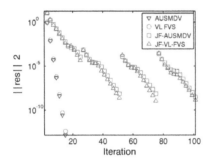

FIGURE 6.7: Effect of discretization of preconditioner.

Chapter 7

The final schemes

Putting the methodology from the last chapters together, we obtain two prototype schemes, a Rosenbrock type scheme and a DIRK type scheme. In the Rosenbrock type schemes, at each stage of the time integration method, a linear system has to be solved, where the matrix is frozen during a time step, whereas the right-hand sides are changing. On the other hand, for the SDIRK type scheme, at each stage of the time integration method, a nonlinear system has to be solved, which means that we obtain a sequence of linear systems with slowly varying right-hand sides and matrices.

All of these schemes need tolerances, termination criteria, and parameters. It is imperative to choose these such that accurate schemes are obtained, but it is also important to avoid oversolving and thus obtain efficient schemes. In the CFD context, this has been discussed in [155, 45] for FV schemes and [150] for DG schemes. We now collect the results documented so far in this book and present choices that lead to efficient schemes. The core problem here is to find good values for parameters where the existing mathematical results are not guidance enough. In particular, good parameters are problem-dependent and different parameters are connected via feedback loops. As an additional goal, there should be as few user-defined parameters as possible. At best, we would end up with just one parameter, namely the tolerance for the time integration error, from which everything else would be determined.

As a further problem, it is difficult to make fair efficiency comparisons of different schemes, because a lot depends on the actual implementation, the test case chosen, and also the tolerance we are interested in. It is clear that it is not sufficient to look at the number of Newton iterations needed or the number of time steps chosen. In particular, time step rejections due to a failure in inner convergence change the picture. Also, the impact of a preconditioner cannot be measured in how much it reduces the number of linear iterations, since we have to consider the setup and application cost as well.

For these reasons, we consider two measures of efficiency, namely the total CPU time and the total number of matrix vector multiplications, which would be equal to the total number of GMRES iterations. Here, total means that we measure this for a complete time integration process from t_0 till t_{end}. The latter indicator is independent of the implementation and of the machine, but does not take into account all the overhead from the Newton iteration, setting up the preconditioner, etc. Furthermore, since GMRES iterations become more costly during one GMRES run, the same number of GMRES iterations for two runs does not mean the same when in a different context of Newton iterations. On the other hand, the CPU time takes these overheads into account, but depends on the machine, the implementation, and what else is going on on the machine at a certain point. Nevertheless, the two together give a good assessment of efficiency.

Here, we will run the numerical tests using two codes. The first is the in-house code TAU_2D. This is a cell-vertex finite volume method with linear reconstruction using the Barth-Jespersen limiter. It uses a primary grid consisting of triangles and quadrilaterals. The boundary conditions are implemented by prescribing fluxes. If not otherwise mentioned, the flux function is AUSMDV.

7.1   DIRK scheme

For a DIRK scheme that uses a Newton-Krylov method for the solution of the appearing nonlinear systems, there are three loops where termination criterions and controls need to be defined: The time integration scheme, the nonlinear solver and the linear solver. In the time integration scheme with error tolerance τ, we have the embedded error estimation based on (4.32) and (4.37). The error estimator might lead to a rejection of time steps, but as discussed, this is not necessary. For the method to solve the nonlinear system, termination is tested using either (5.8) or (5.9) with a tolerance chosen to be $\tau/5$.

As initial guess in Newton's method, we use the last stage value and within GMRES, the zero vector.

As shown in section 5.11, if the time step is chosen too large, Newton's method diverges and the same is true of other methods. Therefore, it is imperative to reject a time step if a tolerance test was failed or if a maximal number of iterations has been reached. In the code SUNDIALS [134], the time step is repeated with $\Delta t_n/4$. Of course, the choice of division by 4 is somewhat arbitrary, but since the original time step was chosen by the time error estimator, a possible decrease by 2 will lead to an increase by the factor $f_{\max}$ in the next step, which is typically chosen to be 2. Therefore, it often happens that in this way, the next time step is rejected again due to the same problems in the nonlinear solver. A division by 4 leads to less overall rejections and thus to a more efficient scheme.

As an additional tweak, we terminate the computation if the time step size is below a certain threshold to avoid a computation stalling with ever tinier time steps.

The linear systems are then solved using a preconditioned Jacobian-free GMRES scheme, where the tolerance is chosen by the Eisenstat-Walker strategy, resulting in a second-order convergent inexact Newton scheme. This was shown to be the fastest option in section 5.11. If the approximate solution of the linear system still fails the tolerance test after a maximal number of iterations, nothing is done except performing another Newton step.

Another important question is the maximal number of Newton steps allowed before a time step is rejected. To get an idea of a reasonable choice, we use TAU_2D with ESDIRK4, ILU preconditioning and the wind turbine problem A.3 with 10 s of simulation time for different tolerances. As initial time step size, $\Delta t = 7.23349 \cdot 10e - 4$ is chosen, corresponding to a CFL number of 10. The results are shown in Table 7.1.

#Newton Steps	10^{-1}		10^{-2}		10^{-3}	
	Iter.	CPU[s]	Iter.	CPU[s]	Iter.	CPU[s]
15	62,784	24,819	134,365	50,891	139,568	51,139
20	113,782	37,662	97,964	32,675	140,542	45,903
25	114,526	37,373	116,694	36,887	158,018	49,133
30	96,654	28,907	124,390	38,064	157,818	48,449
35	122,387	35,905	177,412	53,714	270,918	80,036
40	139,973	40,201	173,602	49,655	205,378	61,044
45	204,970	57,619	211,810	59,471	206,275	61,197
50	219,252	62,589	279,497	78,129	272,123	79,489

TABLE 7.1: Effect of different maximal number of Newton iterations on time-adaptive ESDIRK4 calculation.

First of all we can see that there is a large variance in performance with regards to this parameter and that the feedback loop of rejecting a time step when the Newton loop does not terminate is highly nonlinear. This is because choosing a large number leads to very large time steps with nonlinear systems that take overly long to solve, whereas a small number in a way leads to a bound on the maximal possible time step. Nevertheless, there are two trends visible. First, allowing a too large maximal number of Newton iterations decreases performance. In addition to the reason just mentioned, this is because at some point, more Newton steps only increase the amount of time spent in diverging Newton iterations. Second, setting this too low is also not a good idea, because time step rejections are very inefficient. We need rejections in case of a diverging Newton methods, but if the maximal number of iterations is too low, this implies that convergent Newton iterations will be cut off causing smaller time steps than necessary. Thus, a number of roughly 30 maximal Newton iterations seems to be reasonable.

All in all, we obtain the following algorithm:

- Given error tolerance τ, initial time t_0, and time step size Δt_0, the number of stages s

- For $n = 0, 1, ...$ do

  - For $i = 1, ..., s$

* Compute new right-hand side
* Compute initial guess for Newton scheme (see section 5.7.4)
* For $k = 0, 1, \ldots$ until either (5.8) or (5.9) with tolerance $\tau/5$ is satisfied or
  $k =$ MAX_NEWTON_ITER
    · Determine relative tolerance for linear system solve by (5.57).
    · Solve linear system using preconditioned GMRES with initial guess Zero.

– If MAX_NEWTON_ITER has been reached, but the tolerance test has not been passed, repeat time step with $\Delta t_n = \Delta t_n/4$.
– Estimate local error and compute new time step size Δt_{n+1} using (4.32) and (4.37).
– If $\Delta t_{n+1} < 10^{-20}$, ABORT computation.
– $t_{n+1} = t_n + \Delta t_n$

7.2 Rosenbrock scheme

For Rosenbrock schemes, the algorithm becomes easier, because there is no Newton scheme and thus one loop less. As initial guess in the Krylov subspace method, we still use the zero vector, since the unknown in (4.30) is a sum of stage derivatives, which can be expected to be small. We still have to define tolerances for the linear systems. Furthermore, similar to the control when the nonlinear solver does not terminate in the SDIRK case, we need to handle the situation when the linear solver does not terminate. Another problem is that with the Rosenbrock methods considered here, it may happen that a result obtained in between or after the time step is unphysical. If this happens, meaning that the norm of the new function evaluation is NaN, we repeat the time step with $\Delta t_n/4$, as we do when we encounter this in Newton's method.

Regarding the choice of solver, we have to keep in mind that there is no Newton scheme involved and therefore, the choice of scheme has to be reassessed. To this end, we use the wind turbine test case A.3 for 10 s of simulation time and ROS34PW2, as well as RODASP with an initial time step of $\Delta t = 3.61674 \cdot 10e-4$. The linear systems are solved using ILU preconditioned GMRES up to a tolerance of TOL/100. This was demonstrated to give good stability and accuracy in [45]. The total number of GMRES iterations and the CPU times can be seen in Table 7.2.

First of all, the schemes with a tolerance of 10^{-1} do not converge in the sense that at some time t, the time steps start to get rejected due to NaN appearing and the time integration does not progress anymore. To check if this

Tol.		ROS34PW2		RODASP	
		1st ord. Jac.	Jac. free	1st ord. Jac.	Jac. free
10^{-1}	Time of stall	200.157	202.157	200.783	205.884
10^{-2}	Iter.	353,421	73,750	368,400	178,361
	CPU in s	81,159	21,467	95,552	52,503
10^{-3}	Iter.	427,484	105,647	Div.	240,744
	CPU in s	105,707	32,923		78,162

TABLE 7.2: Efficiency of ROS34PW2 and RODASP if the linear systems are solved based on a first-order Jacobian or using the Jacobian-free GMRES; linear systems are solved up to TOL/100. For the coarsest tolerance, the time of stall of the time integration scheme is mentioned instead.

is due to the linear systems not being solved accurately enough, we repeat the runs and solve the linear systems up to machine accuracy. For ROS34PW2 and the scheme with a first order Jacobian, this leads to an even earlier stall at $t = 200.311s$, whereas the scheme finishes when using Jacobian-free GMRES after 19,266 s of CPU time needing 73,887 GMRES iterations. In the case of RODASP as time integration scheme, solving the linear systems more accurately does not help.

Apparently, we are facing a stability problem, since the main difference resulting from the smaller tolerances are the smaller time steps. This could be a loss of A-stability, which occurs in a ROW-method if the approximation of the Jacobian used is too far away from the Jacobian or some additional nonlinear stability problem. Note that in the study on incompressible flows by John and Rang [148], a Jacobian is computed and kept fix during a time step.

Table 7.2 furthermore shows that the Jacobian-free scheme is significantly faster than the scheme with first-order Jacobian. This deserves further comment. The main reason the JFNK method was faster than the scheme with first-order Jacobian in the DIRK context was that the first one achieves second-order convergence with a large radius of convergence, whereas the latter one is only first-order convergent with a small radius of convergence. Both of these points are irrelevant for the Rosenbrock scheme. Therefore, two different aspects should come to the fore: The finite difference approximation (5.60) is more expensive than a matrix vector product and GMRES with matrix has a faster convergence behavior. Indeed, after about 10 time steps of ROS34PW2 with $TOL = 10^{-2}$, the iteration count of the scheme with first-order Jacobian is 326, whereas it is 1643 for the Jacobian-free scheme.

However, since the scheme with first-order Jacobian solves the wrong linear system at each stage, it incurs larger time integration errors, leading to an average time step of about 10^{-3} and needing 7847 time steps in total to reach the final time. Conversely, the scheme with the Jacobian-free matrix-vector multiplication allows an average time step of about 10^{-1} and needs only 124 time steps to reach the final time. This implies that the first-order Jacobian is not a good approximation of the second-order Jacobian.

Finally, we obtain the following flow diagram for the Rosenbrock type schemes:

- Given error tolerance τ, initial time t_0, and time step size Δt_0, the number of stages s

- For $n = 0, 1, \ldots$ do

  - For $i = 1, \ldots, s$

    * Compute new right-hand side,
    * Solve linear system using a preconditioned Krylov subspace method with relative tolerance $\tau/5$ and initial guess Zero.

  - If $\|\mathbf{f}(\mathbf{u}_{n+1})\| =$ NaN, repeat time step with $\Delta t_n = \Delta t_n/4$.

  - Estimate local error and compute new time step size Δt_{n+1} using (4.33) and (4.37).

  - If $\Delta t_{n+1} < 10^{-20}$, ABORT computation.

  - $t_{n+1} = t_n + \Delta t_n$

7.3   Parallelization

The obvious way of performing parallel computations in the context of the method of lines is to divide the spatial domain into several parts and to assign the computations involving variables from each part to one core. This should be done in a way that minimizes necessary computations between domains. Several ways of obtaining such partitions exist and software that does so can be obtained freely, for example the parMETIS-package [153] which is used here.

Thus we use shared memory parallelization and this is done using the MPI standard, with either the MPICH [3] or the openMPI [272] implementation. The parallel computations considered here are not on massively parallel cluster or on GPUs but on standard multicore architectures.

7.4   Efficiency of finite volume schemes

To test finite volume schemes, we use TAU_2D. In section 5.11, we compared different variants of Newton's method and saw that the JFNK method

Tolerance		None	Jacobi	ILU
10^{-1}	Iter.	319,089	473,692	96,654
	CPU in s	55,580	90,981	28,701
10^{-2}	Iter.	586,013	576,161	124,390
	CPU in s	102,044	111,775	39,056
10^{-3}	Iter.	833,694	642,808	157,818
	CPU in s	142,606	124,901	48,419

TABLE 7.3: CPU times and iteration numbers for the ESDIRK4 scheme for different preconditioners

with the Eisenstat-Walker strategy to determine the tolerances in the GM-RES method was by far the most efficient. This does not take into account the case of Rosenbrock schemes. We now look at the different preconditioners available, where the preconditioner is recomputed every 30 time steps. To this end, We use the time-adaptive ESDIRK4 scheme with 30 maximal Newton iterations and a starting time step of $\Delta t = 7.23349 \cdot 10e - 4$, which corresponds to CFL=10. As a test case, we consider again the wind turbine problem A.3 for a time interval of 10 s. The results can be seen in Table 7.3. As can be seen, ILU reduces the total number of GMRES iterations by a factor of 3 to 6 and is two or three times faster than the code without preconditioning. Block Jacobi is only beneficial for larger tolerances and not very powerful, which means that Jacobi is not a worthwhile preconditioner for finite volume schemes.

Tolerance		SDIRK2	SDIRK3	ESDIRK3	ESDIRK4
10^{-1}	Iter.	72,707	57,703	68,598	96,654
	CPU in s	21,593	19,041	20,736	28,576
10^{-2}	Iter.	54,622	61,042	89,983	124,390
	CPU in s	17,367	21,699	28,377	38,064
		ROS34PW2	RODASP		
10^{-2}	Iter.	73,750	178,361		
	CPU in s	21,467	52,503		

TABLE 7.4: CPU times and iteration numbers for different time integration schemes

We now consider different time integration schemes. The DIRK schemes all use Jacobian-free GMRES, the Eisenstat-Walker strategy, a maximal number of 30 Newton steps and ILU preconditioning. Furthermore, we consider RODASP and ROS34PW2 with Jacobian-free GMRES, which were shown not to work for the coarse tolerance previously in this chapter.

First, we consider the wind turbine test case for a time interval of 10 s and different tolerances. In Table 7.4, we can see the total number of GM-RES iterations and the CPU times in seconds for the various schemes. The fastest is SDIRK2, then comes SDIRK3, ROS34PW2, ESDIRK3, and ES-DIRK4, the worst scheme is RODASP. On the example of SDIRK2, we can see the phenomenon of computational instability [254], meaning that for adaptive schemes, smaller errors not necessarily lead to larger computational cost. In this particular case, SDIRK2 reduces the time step to 10^{-18} by 28 failures to terminate Newton in a row. The reason ESDIRK4 performs so badly is that the time steps chosen are so huge that the nonlinear systems cannot be solved anymore, leading to frequent rejections. The explicit method RK2 needs 40,221 s of CPU time.

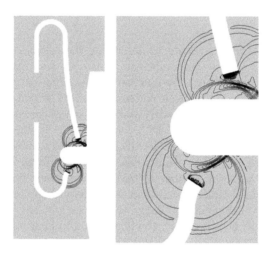

FIGURE 7.1: Pressure isolines.

We now look at a problem motivated from gas quenching [299], which is discussed in detail in section 8.7.1. In practice, high pressured air would be blown from two tubes at a flanged shaft in a cooling process. Here, we follow [45] and instead choose a low Mach number of 0.01 at the outlets of the tubes, to examine the effect of a different source of stiffness. The Reynolds number is 10,000. Regarding initial conditions, the initial velocity is 0, the density 1.2, and the temperature of the gas is 300 K. Thus, the flow is driven by the flow from the tubes, with it being deflected from the flanged shaft after an initial phase. The computation runs until $2 \cdot 1e - 4$ s of real time, which is after the flow of air has been deflected at the shaft.

As a flux function, we use L^2Roe [217], which was discussed in section 3.3.2. The maximal dimension of the Krylov subspace is 40, with the ILU decomposition being updated every 25 time steps. The resulting linear systems are of dimension $568,208$. As initial time step size, we choose $\Delta t = 3.27e - 8$, which corresponds to a CFL number of 10. To compute errors, a reference solution was obtained using RODASP with a tolerance of 10E-6. This solution is depicted in Figure 7.1.

Furthermore, tolerance scaling is employed. Hereby, tolerance is scaled (κ)

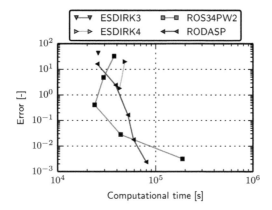

FIGURE 7.2: Adaptive time stepping: Computational efficiency for different time integration schemes for the cooling of a flanged shaft test case. (Credit: D. Blom. From [45], CC-by-SA 4.0 (https://creativecommons.org/licenses/by-sa/4.0/). No changes were made.)

by 1/100 for RODASP, 1/1000 for ROS34PW2 and SDIRK2 and 1/10000 for ESDIRK3 and ESDIRK4, in all cases $\xi = 1$.

In this case the results can be seen in Figure 7.2, only the Rosenbrock schemes were able to provide a solution for all tolerances. With three exceptions at large tolerances, the ESDIRK schemes and SDIRK2 ended up at around $t = 1.7 \cdot 1e - 4$s in a situation where a right-hand side is evaluated with NaN, causing the time step to be repeated with reduced time step size. However, this was repeated again and again, resulting in a stall of the computations with the time step converging to zero. The cause of this is not entirely clear and a reduction of the initial time step did not solve this problem.

Comparing ROS34PW2 and RODASP, we can see that ROS34PW2 performs better for most tolerances, but loses out for smaller tolerances due to the lower order.

7.5 Efficiency of Discontinuous Galerkin schemes

7.5.1 Polymorphic modal-nodal DG

We first consider the polymorphic modal-nodal method described in section 3.7.1 with different preconditioners. The results are taken from [34]. We use a fixed time integration scheme, namely ESDIRK4, which has proven to be a very good DIRK method. After that, we compare different time integration schemes for the preconditioner found best by the first tests.

As a test case, we use vortex shedding behind a 3D sphere A.4, the code is HALO3D, developed at the University of Stuttgart. The relative and absolute error tolerance in the time integration scheme is 10^{-3} and the nonlinear systems are solved using the Jacobian-Free inexact Newton scheme with the Eisenstat-Walker strategy and GMRES. Since storage is a problem, we allow a maximal Krylov subspace dimension of 20.

For the first time step, we chose $\Delta t = 0.0065$, from then on the steps are determined by the time-adaptive algorithm. This time step size was chosen such that the resulting embedded error estimate is between 0.9 and 1, leading to an accepted time step. Then, we perform the computations on a time interval of 30 s. The initial solution and the result at the end when using ESDIRK4 time integration with a tolerance of 10^{-3} are shown in Figure A.8, where we see isosurfaces of lambda2$=-10^{-4}$, a common vortex identifier [147]. Note that there is no optical difference between the results for ESDIRK4 and the LTSRKCK procedure described in section 4.9.2.

Preconditioner	Iter.	CPU in s	Memory in GB
None	218,754	590,968	16.8
Jacobi	23,369	90,004	17.4
ROBO-SGS-0	18,170	77,316	17.4
ROBO-SGS-1	14,639	66,051	19.0
ROBO-SGS-2	13,446	76,450	24.2
ROBO-SGS-3	13,077	87,115	32.1
ROBO-SGS-4	13,061	100,112	43.8
ILU	12,767	108,031	43.8
ILU-CSC	11,609	127,529	43.8

TABLE 7.5: Efficiency of preconditioners for DG

We now compare the use of different preconditioners in Table 7.5. The computations were performed in parallel using eight CPUs. There, the preconditioners are ordered by how powerful we expect them to be, since Jacobi can be seen as the most extreme variant of ROBO-SGS where no information from neighboring cells is taken into account. Thus, as expected from Figure 6.5, the total number of GMRES iterations decrease from no preconditioning to Jacobi to ROBO-SGS-0 up to ROBO-SGS-4, followed by ILU and ILU-CSC. The first thing that can be seen is that by contrast to the finite volume case, Jacobi is a powerful preconditioner. Regarding ROBO-SGS, the decrease is strongest for the first two orders, namely when the constant and the linear part of the solution are taken into account. Consequently, the method of this class with the minimal CPU time is ROBO-SGS-1. Regarding storage, the increase in storage is nonlinear with increasing orders, where ROBO-SGS-4 takes into account the full offdiagonal block and thus needs the same amount of storage as ILU, which is more than double as much as ROBO-SGS-1. This demonstrates again the huge size of the blocks of the Jacobian in a DG scheme and why it is so important to find a way of treating these in an efficient way. Furthermore, it explains why Jacobi is actually a good preconditioner.

When comparing ROBO-SGS to ILU, we see that both ILU and ILU with the coarse scale correction are more powerful preconditioners with ILU-CSC being the overall best preconditioner. However, this does not pay in terms of CPU time and thus the overall best preconditioner is ROBO-SGS-1.

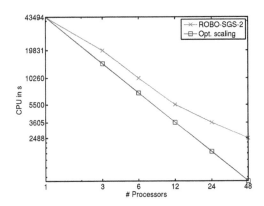

FIGURE 7.3: Parallel efficiency of ROBO-SGS-1.

Now, we demonstrate the strong parallel scaling of ROBO-SGS-1, which means the scaling of solution time with the number of processors for a fixed problem size. To this end, we use the initial conditions from above, but compute only until $t_{end} = 133$ and use a fourth-order method. This saves storage and allows to use the Twelve-Quad-Core machine from the finite volume part. As can be seen in Figure 7.3, the scaling trails off for more than 12 processors. This is roughly the number of processors where the number of unknowns per core drops below 50,000, since we have 740,000 unknowns for this problem. Thus, this behavior is not unexpected.

Scheme	Iter.	CPU in s	Memory in GB
LSERK4	-	346,745	16.3
RKCK	-	80,889	22.9
SDIRK3	22,223	110,844	19.0
ESDIRK3	14,724	73,798	19.0
ESDIRK4	14,639	66,051	19.0
ROS34PW2	60,449	239,869	19.0

TABLE 7.6: Efficiency of time integration schemes for modal-nodal DG.

Finally, we compare different implicit time integration schemes when using ROBO-SGS-1 as a preconditioner with two explicit methods. The results are shown in Table 7.6. As can be seen, the explicit scheme LSERK4 is by far the slowest with about a factor of 5 between ESDIRK4 and LSERK4. Thus, even when considering that a fourth-order explicit scheme is probably overkill for this problem, the message is that explicit RK methods are significantly slower than implicit methods for this test case. Furthermore, the local-time stepping RKCK scheme is about 20% slower than ESDIRK4. Note that for this test case, the time step of the explicit scheme is constrained by the CFL condition and not the DFL condition. Therefore, it can be expected that for higher Reynolds number and finer discretizations at the boundary, the difference in efficiency is even more significant.

When comparing the different implicit schemes, we can see that SDIRK3 is significantly slower than the ESDIRK methods, whereas ESDIRK3 is competitive with ESDIRK4. The worst scheme is the Rosenbrock method, which chooses time steps that are significantly smaller than those of the DIRK methods, which is in line with the experiences from the finite volume case.

7.5.2   DG-SEM

Finally, we consider the second type of DG schemes discussed in the discretization chapter, namely DG-SEM (see 3.7.2). To this end, we consider vortex shedding behind a cylinder in three dimensions at Reynolds number 500 and solve this using the code Strukti, developed at the University of Stuttgart. The result after 1 second of real time for a sixth-order scheme on an $8 \times 8 \times 1$ grid can be seen in Figure 7.4.

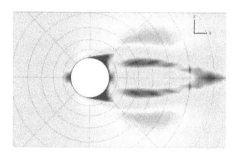

FIGURE 7.4: $|\mathbf{v}|$ around a cylinder at Re 500 after 1 second real time.

To compare efficiency, we use ESDIRK3 with ILU preconditioning and classical Runge-Kutta. The implicit method turns out to be three times slower than the explicit one for a parallel run on eight processors, even though the time step is 50 times higher. The major difference to the modal-nodal scheme is that due to the nonhierarchical basis, ROBO-SGS cannot be applied. Since the size of the blocks for this scheme is even larger than for the modal-nodal scheme, the class of problems for which implicit schemes are more efficient than explicit ones becomes smaller. As an example, higher Reynolds numbers would require a finer resolution, making the implicit scheme more competitive. Nevertheless, more research needs to be put into finding efficient preconditioners for this type of schemes. Here, multigrid methods are quite promising.

Chapter 8

Thermal Fluid Structure Interaction

One example where the computation of unsteady flows often plays a crucial role is fluid-structure interaction. This can happen in different ways. First of all, aerodynamic forces can lead to structural displacement, leading to different forces and so on. The prime example for this are aeroelastic problems, for example flutter of airfoils or bridges or the performance of wind turbines [89]. Second, heat flux from a fluid can lead to temperature changes in a structure, leading to different heat fluxes and so on.

This is called thermal coupling or thermal fluid structure interaction. Examples for this are cooling of gas-turbine blades, thermal anti-icing systems of airplanes [51] or supersonic reentry of vehicles from space [194, 133]. Another application, which will serve as the main motivating example, is quenching, an industrial heat treatment of metal workpieces. There, the desired material properties are achieved by rapid cooling, which causes solid phase changes, allowing to create graded materials with precisely defined properties.

For the solution of a coupled problem, two general approaches can be distinguished. In a partitioned or staggered approach [89], different codes for the subproblems are used and the coupling is done by a main coupling program which calls interface functions of the other codes. This allows to use existing software for each subproblem, in contrast to a monolithic approach, where a new code is tailored for the coupled equations. In the spirit of this book to use flexible components, while still obtaining efficient high order results, we will follow a partitioned approach. We will now explain how the framework of time-adaptive higher-order implicit methods described so far can be used for efficient solution of thermal coupling problems.

8.1 Gas quenching

Gas quenching recently received a lot of industrial and scientific interest [299, 129]. In contrast to liquid quenching, this relatively new process has the advantage of minimal environmental impact because of non-toxic quenching media and clean products like air [263]. Furthermore, it is possible to control the cooling locally and temporally for best product properties and to minimize distortion by means of adapted jet fields (see [245]).

To exploit the multiple advantages of gas quenching, the application of computational fluid dynamics has proved essential [9, 263, 178]. Thus, we consider the coupling of the compressible Navier-Stokes equations as a model for air, along a nonmoving boundary with a nonlinear heat equation as a model for the temperature distribution in steel.

8.2   The mathematical model

The basic setting we are in is that on a domain $\Omega_1 \subset \mathbb{R}^d$ the physics is described by a fluid model, whereas on a domain $\Omega_2 \subset \mathbb{R}^d$, a different model describing a structure is used. A model consists of partial differential equations and boundary conditions, as well as possibly additional algebraic equations. The two domains are almost disjoint where they are connected by an interface. The part of the interface where the fluid and the structure are supposed to interact is called the coupling interface $\Gamma \subset \Omega_1 \cup \Omega_2$ (see Figure 8.1). Note that Γ might be a true subset of the intersection, because the structure could be insulated. At the coupling interface Γ, coupling conditions are prescribed that model the interaction between fluid and structure. For the thermal coupling problem, these conditions are that temperature and the normal component of the heat flux are continuous across the interface.

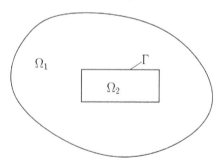

FIGURE 8.1: Illustration of an FSI problem. The domains Ω_1 and Ω_2 meet at an interface, part of that is the coupling interface Γ. (Credit: V. Straub.)

We model the fluid using the time-dependent RANS equations (2.34) with the Spallart-Allmaras turbulence model, written in compact form as

$$\mathbf{u}_t + \nabla \cdot \mathbf{f}^c(\mathbf{u}) = \nabla \cdot \mathbf{f}^v(\mathbf{u}, \nabla \mathbf{u}). \tag{8.1}$$

Additionally, we prescribe appropriate boundary conditions at all parts of the boundary of Ω_1 except Γ.

Regarding the structure model, heat conduction is derived from Fourier's law, but depending on the application, additional equations for thermomechanical effects could and should be included [128]. Here, we will consider heat conduction only. For steel, we have temperature-dependent and highly nonlinear specific heat capacity c_D, heat conductivity λ and emissivity.

$$\rho(\mathbf{x})c_p(\Theta)\frac{d}{dt}\Theta(\mathbf{x},t) = -\nabla \cdot \mathbf{q}(\mathbf{x},t), \tag{8.2}$$

where

$$\mathbf{q}_s(\mathbf{x},t) = -\lambda(\Theta)\nabla\Theta(\mathbf{x},t)$$

denotes the heat flux vector. An empirical model for the steel 51CrV4 was suggested in [228]. The coefficient functions are

$$\lambda(\Theta) = 40.1 + 0.05\Theta - 0.0001\Theta^2 + 4.9 \cdot 10^{-8}\Theta^3 \tag{8.3}$$

and

$$c_p(\Theta) = -10\ln\left(\frac{e^{-c_{p1}(\Theta)/10} + e^{-c_{p2}(\Theta)/10}}{2}\right) \tag{8.4}$$

with

$$c_{p1}(\Theta) = 34.2e^{0.0026\Theta} + 421.15 \tag{8.5}$$

and

$$c_{p2}(\Theta) = 956.5e^{-0.012(\Theta-900)} + 0.45\Theta. \tag{8.6}$$

For the mass density one has $\rho = 7836$ kg/m^3.

On the boundary, we have Neumann conditions $\mathbf{q}(\mathbf{x},t) \cdot \mathbf{n}(\mathbf{x}) = q_b(\mathbf{x},t)$.

Finally, initial conditions for both domains, $\Theta(\mathbf{x},t_0) = \Theta_0(\mathbf{x})$, $\mathbf{u}(\mathbf{x},t_0) = \mathbf{u}_0(\mathbf{x})$ are required.

As mentioned at the beginning of this section, the coupling conditions are that temperature and the normal component of the heat flux are continuous across the interface, i.e.,

$$T(\mathbf{x},t) = \Theta(\mathbf{x},t), \quad \mathbf{x} \in \Gamma, \tag{8.7}$$

where T is the fluid temperature and

$$\mathbf{q}_f(\mathbf{x},t) \cdot \mathbf{n}(x) = \mathbf{q}_s(\mathbf{x},t) \cdot \mathbf{n}(x), \quad \mathbf{x} \in \Gamma, \tag{8.8}$$

where $\mathbf{q}_f$ is the heat flux from the fluid.

8.3   Space discretization

Following the partitioned coupling approach, we discretize the two models separately in space. For the fluid, we use a finite volume method, leading to

$$\frac{d}{dt}\underline{\mathbf{u}} + \underline{\mathbf{h}}(\underline{\mathbf{u}}, \boldsymbol{\Theta}_\Gamma) = \mathbf{0}, \tag{8.9}$$

where the dependence on the discrete interface structures is made explicit, here denoted by $\boldsymbol{\Theta}_\Gamma$.

Regarding structural mechanics, the use of finite element methods is ubiquous. Therefore, we will also follow that approach here and use linear finite elements, leading to the nonlinear equation for all unknowns on Ω_2

$$\mathbf{M}(\boldsymbol{\Theta})\frac{d}{dt}\boldsymbol{\Theta}(t) + \mathbf{K}(\boldsymbol{\Theta})\boldsymbol{\Theta}(t) = \mathbf{q}_b + \bar{\mathbf{q}}(\underline{\mathbf{u}}(t)). \tag{8.10}$$

Here, $\mathbf{M}$ is the heat capacity and $\mathbf{K}$ the heat conductivity matrix. The vector $\boldsymbol{\Theta}$ consists of all discrete temperature unknowns and $\bar{\mathbf{q}}(\underline{\mathbf{u}}(t))$ is the heat flux vector on the coupling interface to the fluid. The discretization of the coupling condition (8.8) is hereby assumed to be a part of (8.10).

8.4   Coupled time integration

A number of articles devoted to the aspect of time integration for fluid-structure interaction problems have been published. In [187], energy conservation for a problem from aeroelasticity is analyzed using the implicit midpoint rule in a monolithic scheme. Already in [17, 67], it is suggested to use an explicit high order RK scheme for both subproblems with data exchange at each stage. Due to the explicit nature, the resulting scheme has severely limited time steps. The order of coupling schemes on moving meshes is analyzed in [116], but only convergence of first order is proved for pth order schemes. Furthermore, the combination of higher order IRK schemes for problems on moving meshes is explored in [288] for the one-dimensional case and in the subsequent paper [289] for 3D calculations. There, so-called explicit first stage, singly diagonally implicit Runge-Kutta schemes (ESDIRK) are employed and higher order in time is demonstrated by numerical results.

Now, as explained before, if the fluid and the solid solver are able to carry out time steps of implicit Euler type, the main coupling program of the FSI procedure can be extended to SDIRK methods very easily, since the main coupling program just has to call the backward Euler routines with specific time step sizes and starting vectors.

If time adaptivity is added, things become slightly more complicated, because in formula (4.32), all stage derivatives are needed. Therefore, these have to be stored by the main coupling program or the subsolvers. We choose the latter; (4.37) is applied locally and the error estimates are reported back to the main coupling program. Furthermore, if the possibility of rejected time steps is taken into account, $\mathbf{y}_n$ has to be stored as well. Accordingly, the possibilities to achieve higher order accurate results are opened. This approach was first suggested in [41].

In the following, it is assumed that the step size Δt_n (or $a_{ii}\Delta t_n$) is prescribed. To show the global procedure at stage i, the starting vector (4.20) of the DIRK-method is decomposed into the fluid and solid variables, $\mathbf{s}_i = \{\mathbf{s}_i^{\mathbf{u}}, \mathbf{S}_i^{\Theta}\}$. According to equations (8.9)–(8.10), the coupled system of equations

$$\mathbf{F}(\underline{\mathbf{u}}_i, \mathbf{\Theta}_i) := \mathbf{u}_i - \mathbf{s}_i^{\mathbf{u}} - \Delta t_n\, a_{ii} \underline{\mathbf{h}}(\underline{\mathbf{u}}_i, \mathbf{\Theta}_i) = \mathbf{0} \tag{8.11}$$

$$\mathbf{T}(\underline{\mathbf{u}}_i, \mathbf{\Theta}_i) := [\mathbf{M}(\mathbf{\Theta}) - \Delta t_n\, a_{ii} \mathbf{K}(\mathbf{\Theta})]\mathbf{\Theta}_i - \mathbf{M}(\mathbf{\Theta})\mathbf{S}_i^{\Theta} - \Delta t_n\, a_{ii} \bar{\mathbf{q}}(\underline{\mathbf{u}}_i) = \mathbf{0} \tag{8.12}$$

has to be solved.

The dependence of the fluid equations $\underline{\mathbf{h}}(\underline{\mathbf{u}}_i, \mathbf{\Theta}_i)$ on the temperature $\mathbf{\Theta}_i$ results from the nodal temperatures of the structure at the interface. This subset is written as $\mathbf{\Theta}_i^{\Gamma}$. Accordingly, the structure equations depend only on the heat flux of the fluid at the coupling interface.

8.5 Dirichlet-Neumann iteration

In each stage i, the coupled system of nonlinear equations (8.11)–(8.12) has to be solved. The standard method is a nonlinear fixed point methods, namely Block-Gauß-Seidel. With our choice of boundary conditions (temperature for the fluid and heat flux for the structure), this is called a Dirichlet-Neumann iteration. The algorithm is

$$\mathbf{F}(\underline{\mathbf{u}}_i^{(\nu+1)}, \mathbf{\Theta}_i^{(\nu)}) = \mathbf{0} \quad \rightsquigarrow \quad \underline{\mathbf{u}}_i^{(\nu+1)} \tag{8.13}$$

$$\mathbf{T}(\underline{\mathbf{u}}_i^{(\nu+1)}, \mathbf{\Theta}_i^{(\nu+1)}) = \mathbf{0} \quad \rightsquigarrow \quad \mathbf{\Theta}_i^{(\nu+1)} \tag{8.14}$$

within the iteration index (ν). The starting values are given by $\underline{\mathbf{u}}_i^{(0)} = \mathbf{s}_i^{\mathbf{u}}$ and $\mathbf{\Theta}_i^{(0)} = \mathbf{s}_i^{\Theta}$. The termination criterion is formulated by the nodal temperatures at the interface of the solid structure

$$\|\mathbf{\Theta}_i^{\Gamma\,(\nu+1)} - \mathbf{\Theta}_i^{\Gamma\,(\nu)}\| \leq TOL_{FIX}. \tag{8.15}$$

The tolerance in the time-adaptive case is again given by the tolerance for the time integration, divided by 5. If the iteration is stopped after one step, e.g., only one computation of (8.13)–(8.14) is carried out, this is also referred to as loose coupling, whereas iterating until the termination criterion is fulfilled corresponds to strong coupling. Note that the subsolvers involve in themselves iterative solvers. The tolerance there would be given by the tolerance of the fixed point iteration, or smaller. In case the subsolvers both solve linear problems, care has to be taken with regards to the type of tolerance criterion employed [27].

The convergence rate of the Dirichlet-Neumann analysis for coupled linear heat equations has been analyzed in [106, 130, 40, 205, 110]. As is not uncommon for coupled problems, it depends on ratios of material parameters, but also on the discretizations that are coupled. As shown in [40], for coupled FEM discretizations in one dimension, the asymptotic convergence rate for Δt large compared to Δx^2 is the quotient of heat conductivities, $\frac{\lambda_1}{\lambda_2}$. In the limit Δt to zero, we surprisingly do not obtain a spectral radius of the iteration matrix of zero, but of the ratio of the products of densities time specific heat capacity. Thus, for certain couplings of materials, the convergence rate can go down when decreasing the time step. On the other hand, for the coupling of a finite element method and a finite volume method, the spectral radius of the iteration matrix does go to zero for Δt to zero [205]. For Δt large, we again obtain the quotient of heat conductivities. This combination of discretizations is standard and the one discussed in this chapter. As turns out, the aspect ratio plays a crucial role as well and the convergence rate is proportional to it. This was also demonstrated numerically.

From the results above, it follows that temperature (the Dirichlet condition) has to be prescribed for the equation with smaller heat conductivity, here the fluid, and heat flux is prescribed for the structure. Choosing the conditions the other way round leads to an unstable scheme. It also follows that because the quotient of heat conductivities of air and steel is tiny, the convergence rates are expected to be very small, which indeed can be observed in practice [36, 205].

Various methods have been proposed to increase the convergence speed of the fixed point iteration by decreasing the interface error between subsequent steps, for example Relaxation [173, 165], Interface-GMRES [201], or ROM-coupling [292]. Relaxation means that after the fixed point iterate is computed, a relaxation step is added:

$$\tilde{\Theta}_i^{\Gamma\,(\nu+1)} = \omega \Theta_i^{\Gamma\,(\nu+1)} + (1-\omega)\Theta_i^{\Gamma\,(\nu)}.$$

Several strategies exist to compute the relaxation parameter ω, in particular Aitken relaxation, which amounts to

$$\omega_{\nu+1} = -\omega_\nu \frac{(\mathbf{r}^{\Gamma\,(\nu)})^T(\mathbf{r}^{\Gamma\,(\nu+1)} - \mathbf{r}^{\Gamma\,(\nu)})}{\|\mathbf{r}^{\Gamma\,(\nu+1)} - \mathbf{r}^{\Gamma\,(\nu)}\|^2}.$$

Here, we use the interface residual

$$\mathbf{r}^{\Gamma\,(\nu)} = \Theta_i^{\Gamma\,(\nu+1)} - \Theta_i^{\Gamma\,(\nu)}.$$

8.5.1 Extrapolation from time integration

To find an initial guess for the Dirichlet-Neumann iteration, we suggest to make use of knowledge about the time integration to do extrapolation of the interface temperatures. For SDIRK2, the following formations have been derived in [36].

At the first stage, we have the old time step size Δt_{n-1} with value Θ_{n-1} and the current time step size Δt_n with value Θ_n. We are looking for the value Θ_n^1 at the next stage time $t_n + c_1 \Delta t_n$. Linear extrapolation gives

$$\Theta_n^1 \approx \Theta_n + c_1 \Delta t_n (\Theta_n - \Theta_{n-1})/\Delta t_{n-1} = \left(1 + \frac{c_1 \Delta t_n}{\Delta t_{n-1}}\right) \Theta_n - \frac{c_1 \Delta t_n}{\Delta t_{n-1}} \Theta_{n-1}.$$
(8.16)

Regarding quadratic extrapolation, it is reasonable to choose t_n, t_{n-1}, and the intermediate temperature vector Θ_{n-1}^1 from the previous stage $t_{n-1} + c_1 \Delta t_{n-1}$. This results in

$$\begin{aligned}\Theta_n^1 \approx \quad & \Theta_{n-1} \frac{(c_1 \Delta t_n + (1-c_1)\Delta t_{n-1})c_1 \Delta t_n}{c_1 \Delta t_{n-1}^2} - \Theta_{n-1}^1 \frac{(c_1 \Delta t_n + \Delta t_{n-1})c_1 \Delta t_n}{c_1 \Delta t_{n-1}^2 (1-c_1)} \\ & + \Theta_n \frac{(c_1 \Delta t_n + \Delta t_{n-1})(c_1 \Delta t_n + (1-c_1)\Delta t_{n-1})}{(1-c_1)\Delta t_{n-1}^2}.\end{aligned}$$
(8.17)

At the second stage, we linearly extrapolate Θ_n at t_n and Θ_n^1 at $t_n + c_1 \Delta t$ to obtain

$$\Theta_{n+1} \approx \Theta_n + \Delta t_n (\Theta_n^1 - \Theta_n)/(c_1 \Delta t_n) = \left(1 - \frac{1}{c_1}\right)\Theta_n + \frac{1}{c_1}\Theta_n^1.$$
(8.18)

When applying quadratic extrapolation at the second stage (or at later stages in a scheme with more than 2), it is better to use values from the current time interval. This results in

$$\begin{aligned}\Theta_{n+1} \approx \quad & \Theta_{n-1} \frac{\Delta t_n^2 (1-c_1)}{\Delta t_{n-1}(\Delta t_{n-1} + c_1 \Delta t_n)} - \Theta_n \frac{(\Delta t_{n-1} + \Delta t_n)(1-c_1)\Delta t_n}{\Delta t_{n-1} c_1 \Delta t_n} \\ & + \Theta_n^1 \frac{(\Delta t_{n-1} + \Delta t_n)\Delta t_n}{(c_1 \Delta t_n + \Delta t_{n-1})c_1 \Delta t_n}.\end{aligned}$$
(8.19)

8.6 Alternative solvers

In the method described above, time is synchronized. Another approach would be to define a synchronization time and let the subsolvers do their time integration independently within such a time window. This leads to waveform relaxation methods, where instead of interface data at a time step, (interpolated) time continuous interface functions are exchanged. This is less invasive from the standpoint of partitioned fluid structure interaction. For conjugate heat transfer, different promising variants have been suggested [95, 206, 207].

The method described above works very well for this particular problem. Another approach would be black box methods that work for a wide range of problems. Newton-type methods have for example been considered [189], but they require cross derivatives, which are hard to obtain. They can be combined with the waveform relaxation methods mentioned above, in the form of Quasi-Newton methods. This leads to Quasi-Newton Waveform relaxation methods [239].

8.7   Numerical Results

For the calculations presented here, we used the DLR TAU-Code [104] for the flow and the in-house quadratic finite element code Native of the Institute for Static and Dynamic at the University of Kassel for the solid. TAU uses a cell-centered finite volume method with linear reconstruction and FAS to solve the appearing nonlinear equation systems. The technical difficulty of different programming languages (FORTRAN for Native and C++ for TAU) in the partitioned approach is dealt with a C++-library called Component Template Library (CTL) [188].

8.7.1   Cooling of a flanged shaft

To illustrate a realistic example of gas quenching, the cooling of a flanged shaft by cold high pressured air is used. The complete process consists of the inductive heating of a steel rod, the forming of the hot rod into a flanged shaft, a transport to a cooling unit, and the cooling process. Here, we consider only the cooling, which means that we have a hot flanged shaft that is cooled by cold high-pressured air coming out of small tubes [299]. We perform a two-dimensional cut through the domain and assume symmetry along the vertical axis, resulting in one-half of the flanged shaft and two tubes blowing air at it, (see Figure 8.2). Since the air nozzles are evenly distributed around the flanged shaft, we use an axisymmetric model in the structure. The heat flux from the two-dimensional simulation of the fluid at the boundary of the flanged shaft is impressed axially symmetrical on the structure.

We assume that the air leaves the tube in a straight and uniform way at a Mach number of 1.2. Furthermore, we assume a freestream in x-direction of Mach 0.005. This is mainly to avoid numerical difficulties at Mach 0, but could model a draft in the workshop. The Reynolds number is $Re = 2500$ and the Prandtl number $Pr = 0.72$.

The grid consists of 279,212 cells in the fluid, which is the dual grid of an unstructured grid of quadrilaterals in the boundary layer and triangles in the rest of the domain, and 1997 quadrilateral elements in the structure. It is illustrated in Figure 8.3.

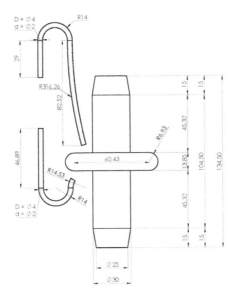

FIGURE 8.2: Sketch of the flanged shaft. (Credit: Forming Technology, University of Kassel.)

To obtain initial conditions for the subsequent tests, we use the following procedure: We define a first set of initial conditions by setting the flow velocity to zero throughout and choose the structure temperatures at the boundary points to be equal to temperatures that have been measured by a thermographic camera. Then, setting the y-axis on the axis of revolution of the flange, we set the temperature at each horizontal slice to the temperature at the corresponding boundary point. Finally, to determine the actual initial conditions, we compute 10^{-5} s of real time using the coupling solver with a fixed time step size of $\Delta t = 10^{-6}s$. This means, that the high-pressured air is coming out of the tubes and the first front has already hit the flanged shaft. This solution is illustrated in Figure 8.4 (left).

Now, we compute 1 second of real time using the time-adaptive algorithm with different tolerances and an initial time step size of $\Delta t = 10^{-6}s$. This small initial step size is necessary to prevent instabilities in the fluid solver. During the course of the computation, the time step size is increased until it is on the order of $\Delta t = 0.1s$, which demonstrates the advantages of the time-adaptive algorithm and reaffirms that it is this algorithm that we need to compare to. In total, the time-adaptive method needs 22, 41, 130, and 850 time steps to reach $t = 1s$ for the different tolerances, compared to the 10^6 steps the fixed time step method would need. The solution at the final time is depicted in Figure 8.4 (right). As can be seen, the stream of cold air is deflected by the shaft.

Finally, we consider extrapolation based on the time integration scheme. In Table 8.1, the total number of iterations for 1 second of real time is shown.

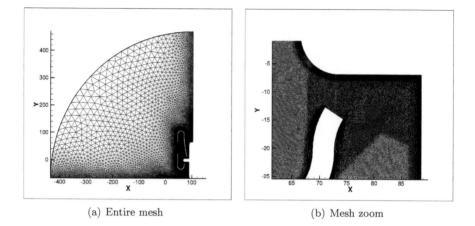

(a) Entire mesh   (b) Mesh zoom

FIGURE 8.3: Grid around flanged shaft. Left: Complete computational domain. Right: Zoom on region around lower tube and shaft. (Credit: From [45], CC-by-SA 4.0 (https://creativecommons.org/licenses/by-sa/4.0/).)

As before, the extrapolation causes a noticeable decrease in the total number of fixed point iterations. The speedup from linear extrapolation is between 18% and 34%, compared to the results obtained without extrapolation.

TOL	none	lin.
10^{-2}	51	42
10^{-3}	126	97
10^{-4}	414	309
10^{-5}	2768	1805

TABLE 8.1: Total number of iterations for 1 second of real time for different extrapolation methods in time.

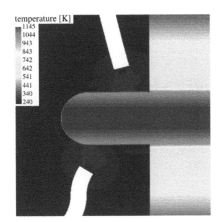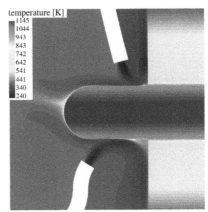

FIGURE 8.4: Temperature distribution in fluid and structure at $t = 0s$ (left) and $t = 1s$ (right). (Credit: T. Gleim.)

Appendix A

Test problems

A.1 Shu-vortex

For the Euler equation, a two-dimensional problem where the exact solution is known is the convection of a two-dimensional vortex in a periodic domain, which is designed to be isentropic [68]. The initial conditions are taken as freestream values $(\rho_\infty, v_{1_\infty}, v_{2_\infty}, T_\infty) = (1, v_{1_\infty}, 0, 1)$, but are perturbed at t_0 by a vortex $(\delta v_1, \delta v_2, \delta T)$ centered at $(\tilde{x}_1, \tilde{x}_2)$, given by the formulas

$$
\begin{aligned}
\delta v_1 &= -\frac{\alpha}{2\pi}(x_2 - \tilde{x}_2)e^{\phi(1-r^2)}, \\
\delta v_2 &= \frac{\alpha}{2\pi}(x_1 - \tilde{x}_1)e^{\phi(1-r^2)}, \\
\delta T &= -\frac{\alpha^2(\gamma - 1)}{16\phi\gamma\pi^2}(x_1 - \tilde{x}_1)e^{2\phi(1-r^2)},
\end{aligned}
\tag{A.1}
$$

where ϕ and α are parameters and r is the euclidian distance of a point (x_1, x_2) to the vortex center, which is set at $(\tilde{x}_1, \tilde{x}_2) = (0, 0)$ in the domain $[-7, 7] \times [-3.5, 3.5]$. The parameter α can be used to tweak the speed of the flow in the vortex, whereas ϕ determines the size. The initial solution can be seen in Figure A.1.

The functions (A.1) do not provide an exact solution of the Navier-Stokes equations. There, we have additional diffusion, which leads to a decay of the solution in time. To test our methods, we use a structured grid with 23,500 cells, which is illustrated in Figure A.2.

A.2 Supersonic flow around a cylinder

The second test case is the two-dimensional flow around a cylinder at Mach 10. The grid consists of 20,994 points, whereby only a quarter of the domain is discretized, and system matrices are of dimension 83976. As initial data, freestream conditions are used. The steady state can be seen in Figure A.3.

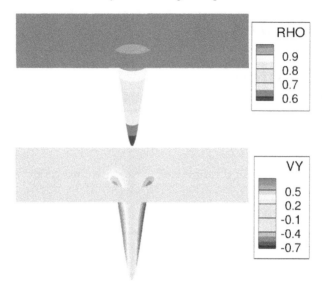

FIGURE A.1: Initial solution of isentropic vortex solution of the Euler equations. Density (left) and v_y (right).

A.3   Wind turbine

The third test case is a two-dimensional flow around the cross section of the wind turbine DUW-96, designed by the wind energy research initiative DUWIND of the TU Delft. The Mach number is 0.12 with an angle of attack of 40° and the Reynolds number 1000. The grid has 24,345 cells and is illustrated in Figure A.4.

To obtain an unsteady test case, we first compute the steady state around this for an Euler flow. When starting from this solution, an immediate vortex shedding starts, which is slow due to the low Reynolds number. The steady state solution and the ones after 30, 60, and 90 s of simulation time can be seen in Figures A.5 and A.6. These were computed using SDIRK2 at a tolerance of 10^{-2}.

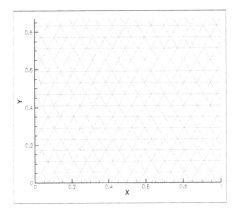

FIGURE A.2: Excerpt of grid for Shu vortex problem.

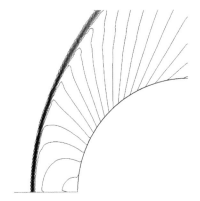

FIGURE A.3: Pressure isolines for a cylinder at Mach 10.

A.4   Vortex shedding behind a sphere

This test case is a three-dimensional test case about vortex shedding behind a sphere at Mach 0.3 and a Reynolds number of 1000. The grid is illustrated in Figure A.7. It consists of 21,128 hexahedral cells.

To obtain initial conditions and start the initial vortex shedding, we begin with free stream data and with a second-order method. Furthermore, to cause an initial disturbance, the boundary of the sphere is chosen to be noncurved at first. In this way, we compute 120 s of time. Then we switch to fourth order, compute another 10 s of time. From this, we start the actual computations

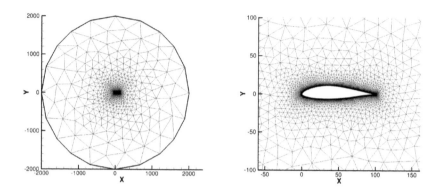

FIGURE A.4: Grid around wind turbine.

with a fifth-order method and a curved representation of the boundary. Thus, we have 739,480 unknowns.

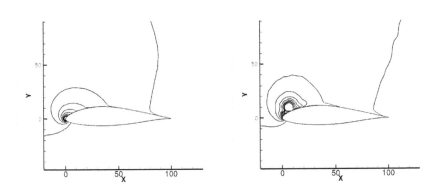

FIGURE A.5: Density isolines for wind turbine problem at $t = 0\,s$ and $t = 30\,s$.

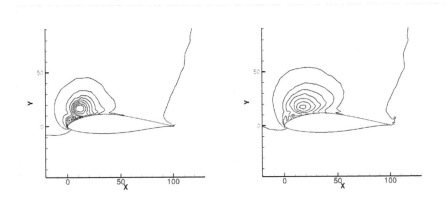

FIGURE A.6: Density isolines for wind turbine problem at $t = 60\,s$ and $t = 90\,s$.

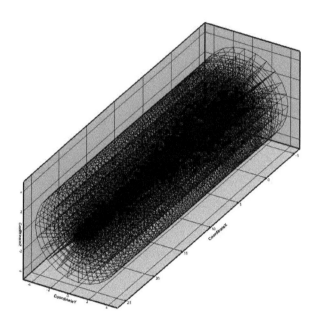

FIGURE A.7: Grid for sphere test case.

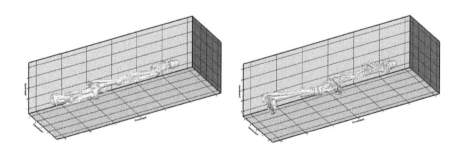

FIGURE A.8: Isosurfaces of lambda-2$=-10^{-4}$ for initial (left) and final solution of sphere problem (right).

Appendix B

Coefficients of time integration methods

$$
\begin{array}{c|cc}
1 & 1 \\
\hline
 & 1/2 & 1/2
\end{array}
\qquad\qquad
\begin{array}{c|ccc}
1 & 1 \\
1/2 & 1/4 & 1/4 \\
\hline
 & 1/6 & 1/6 & 2/3
\end{array}
$$

TABLE B.1: Explicit SSP methods: SSP2 and SSP3

$$
\begin{array}{c|cccc}
0 & 0 & 0 & 0 & 0 \\
1/2 & 1/2 & 0 & 0 & 0 \\
1/2 & 0 & 1/2 & 0 & 0 \\
1 & 0 & 0 & 1 & 0 \\
\hline
 & 1/6 & 1/3 & 1/3 & 1/6
\end{array}
$$

TABLE B.2: Coefficients for RK4

i	a_i	b_i
1	0	$\dfrac{1432997174477}{9575080441755}$
2	$-\dfrac{567301805773}{1357537059087}$	$\dfrac{5161836677717}{13612068292357}$
3	$-\dfrac{2404267990393}{2016746695238}$	$\dfrac{1720146321549}{2090206949498}$
4	$-\dfrac{3550918686646}{2091501179385}$	$\dfrac{3134564353537}{4481467310338}$
5	$-\dfrac{1275806237668}{842570457699}$	$\dfrac{2277821191437}{14882151754819}$

TABLE B.3: LSERK4 coefficients

$\frac{4-\sqrt{6}}{10}$	$\frac{88-7\sqrt{6}}{360}$	$\frac{296-169\sqrt{6}}{1800}$	$\frac{-2+3\sqrt{6}}{225}$
$\frac{4+\sqrt{6}}{10}$	$\frac{296+169\sqrt{6}}{1800}$	$\frac{88+7\sqrt{6}}{360}$	$\frac{-2-3\sqrt{6}}{225}$
1	$\frac{16-\sqrt{6}}{36}$	$\frac{16+\sqrt{6}}{36}$	$1/9$
	$\frac{16-\sqrt{6}}{36}$	$\frac{16+\sqrt{6}}{36}$	$1/9$

1/3	5/12	-1/12
1	3/4	1/4
	3/4	1/4

TABLE B.4: Butcher array for the Radau IIA method of orders 3 and 5.

0	1/2	-1/2
1	1/2	1/2
	1/2	1/2

0	1/6	-1/3	1/6
1/2	1/6	5/12	-1/12
1	1/6	2/3	1/6
	1/6	2/3	1/6

TABLE B.5: Butcher arrays for the Gauß-Lobatto IIIC methods of order 2 and 4

0	1/12	$-\frac{\sqrt{5}}{12}$	$\frac{\sqrt{5}}{12}$	-1/12
$\frac{5-\sqrt{5}}{10}$	1/12	1/4	$\frac{10-7\sqrt{5}}{60}$	$\frac{\sqrt{5}}{60}$
$\frac{5+\sqrt{5}}{10}$	1/12	$\frac{10+7\sqrt{5}}{60}$	1/4	$-\frac{\sqrt{5}}{60}$
1	1/12	5/12	5/12	1/12
	1/12	5/12	5/12	1/12

TABLE B.6: Butcher array for the Gauß-Lobatto IIIC method of order 6

α	α	0
1	$1-\alpha$	α
b_i	$1-\alpha$	α
$\hat{b}_i$	$1-\hat{\alpha}$	$\hat{\alpha}$
$b_i - \hat{b}_i$	$\hat{\alpha}-\alpha$	$\alpha-\hat{\alpha}$

$$\alpha = 1 - \sqrt{2}/2$$
$$\hat{\alpha} = 2 - \tfrac{5}{4}\sqrt{2}$$
$$\alpha - \hat{\alpha} = -1 + \tfrac{3}{4}\sqrt{2}$$

TABLE B.7: Butcher array for the method of Ellsiepen

γ	γ	0	0
δ	$\delta-\gamma$	γ	0
1	α	β	γ
b_i	α	β	γ
$\hat{b}_i$	$\hat{\alpha}$	$\hat{\beta}$	0
$b_i - \hat{b}_i$			γ

$$\alpha = 1.2084966491760101$$
$$\beta = -0.6443631706844691$$
$$\gamma = 0.4358665215084580$$
$$\delta = 0.7179332607542295$$
$$\delta - \gamma = 0.2820667392457705$$
$$\hat{\alpha} = 0.7726301276675511$$
$$\hat{\beta} = 0.2273698723324489$$

TABLE B.8: Butcher array for the method of Cash

0	0	0	0	0
2γ	γ	γ	0	0
3/5	0.257648246066427	-0.093514767574886	γ	0
1	0.187641024346724	-0.595297473576955	0.971789927721772	γ
b_i	0.187641024346724	-0.595297473576955	0.971789927721772	γ
$\hat{b}_i$	0.214740286223389	-0.485162263884939	0.868725002520388	0.401696975141162
$b_i - \hat{b}_i$	-0.027099261876665	-0.110135209692016	0.103064925201385	0.034169546367297

TABLE B.9: Butcher diagram for ESDIRK3. $\gamma = 0.435866521508459$

0	0	0	0	0	0
2γ	γ	γ	0	0	0
83/250	0.137776	-0.055776	γ	0	0
31/50	0.144636866026982	-0.223931907613345	0.449295041586363	γ	0
17/20	0.098258783283565	-0.591544242819670	0.810121053828300	0.283164405707806	γ
1	0.157916295161671	0	0.186758940524001	0.680565295309335	-0.275240530995007
b_i	0.157916295161671	0	0.186758940524001	0.680565295309335	-0.275240530995007
$\hat{b}_i$	0.154711800763212	0	0.189205191660680	0.702045371228922	-0.319187399063579
$b_i - \hat{b}_i$	0.003204494398459	0	-0.002446251136679	-0.021480075919587	0.043946868068572

TABLE B.10: Butcher diagram for ESDIRK4. The last column has been left out for space reasons. $\gamma = 0.25$ and the missing component $\hat{b}_6 = 0.273225035410765$

j \ i	1	2	3	4
1	0.203524508852533	-1.888641733544525	2.596814547851191	0.088302676840802
2	-0.015883484505809	1.29334425996757	-1.625024620129419	0.347563844667657

TABLE B.11: Coefficients b^*_{ij} of the dense output formulas for ESDIRK3

j \ i	1	2	3	4	5	6
1	0.924880883935497	0	0.73854539473069	-2.517224551339271	3.5887741459555	-1.734975873282416
2	-0.766964588773826	0	-0.551786454206689	3.197789846648606	-3.864014676976506	1.984975873282416

TABLE B.12: Coefficients b^*_{ij} of the dense output formulas for ESDIRK4

$\gamma = 0.435866521508459$

$\tilde{a}_{21} = 0.8717330430169180\,1$	$\gamma_{21} = -0.8717330430169180\,1$	$a_{21} = 2$	$c_{21} = -4.5885607205580\,85$
$\tilde{a}_{31} = 0.8445706001536942\,3$	$\gamma_{31} = -0.9033805701304408\,2$	$a_{31} = 1.4192173174557\,65$	$c_{31} = -4.1847604823191\,61$
$\tilde{a}_{32} = -0.1129906423648418\,5$	$\gamma_{32} = 0.0541806723880953\,26$	$a_{32} = -0.2592322116729\,70$	$c_{32} = 0.2851920173554\,96$
$\tilde{a}_{41} = 0$	$\gamma_{41} = 0.2421238070609534\,6$	$a_{41} = 4.1847604823191\,61$	$c_{41} = -6.3681792001283\,6$
$\tilde{a}_{42} = 0$	$\gamma_{42} = -1.2232505083904514\,7$	$a_{42} = -0.2851920173554\,96$	$c_{42} = -6.7956209446683\,7$
$\tilde{a}_{43} = 1$	$\gamma_{43} = 0.545260255335102\,14$	$a_{43} = 2.2942803602790\,42$	$c_{43} = -2.8700986043310\,56$
$b_1 = 0.2421238070609534\,6$	$\hat{b}_1 = 0.3781090314581369$	$m_1 = 4.1847604823191\,60$	$\hat{m}_1 = 3.9070105346711\,93$
$b_2 = -1.2232505083904514\,7$	$\hat{b}_2 = -0.0960422922124231\,78$	$m_2 = -0.2851920173554\,97$	$\hat{m}_2 = 1.1180478778205\,03$
$b_3 = 1.545260255335102$	$\hat{b}_3 = 0.5$	$m_3 = 2.2942803602790\,42$	$\hat{m}_3 = 0.5216502326114\,91$
$b_4 = 0.435866521508459$	$\hat{b}_4 = 0.217933260754229\,5$	$m_4 = 1$	$\hat{m}_4 = 0.5$

TABLE B.13: Coefficients for ROS34PW2

$\gamma = 0.25$

$\tilde{a}_{21} = 0.75$	$\gamma_{21} = -0.75$	$a_{21} = 3.0$	$c_{21} = -12.0$
$\tilde{a}_{31} = 0.0861204008141537\,5$	$\gamma_{31} = -0.1355124008141537\,5$	$a_{31} = 1.8310367934867\,70$	$c_{31} = -8.7917951739470\,78$
$\tilde{a}_{32} = 0.1238795991858462\,5$	$\gamma_{32} = -0.1379915991858462\,5$	$a_{32} = 0.4955183967433\,85$	$c_{32} = -2.2078655869735\,39$
$\tilde{a}_{41} = 0.7740345350732462$	$\gamma_{41} = -1.2560840048950797$	$a_{41} = 2.3043765882692\,655$	$c_{41} = 10.817930568571\,748$
$\tilde{a}_{42} = 0.1492651549508692\,4$	$\gamma_{42} = -0.2501447105064247\,9$	$a_{42} = -0.0524927252457\,437$	$c_{42} = 6.7802760611428\,374$
$\tilde{a}_{43} = -0.2941996904581938\,6$	$\gamma_{43} = 1.2209287154015045$	$a_{43} = -1.1767987618327\,76$	$c_{43} = 19.534859446424\,075$
$\tilde{a}_{51} = 5.308746682646143$	$\gamma_{51} = -7.0731843331420627\,9$	$a_{51} = -7.1704549624232\,04$	$c_{51} = 34.190950067497\,354$
$\tilde{a}_{52} = 1.330892140037269\,3$	$\gamma_{52} = -1.805648697243572\,4$	$a_{52} = -4.7416366714818\,72$	$c_{52} = 15.496711537259\,923$
$\tilde{a}_{53} = -5.3741378116555623$	$\gamma_{53} = 7.743829658571367\,1$	$a_{53} = -16.310026313309\,713$	$c_{53} = 54.747608759641\,373$
$\tilde{a}_{54} = -0.2655010110278499\,9$	$\gamma_{54} = 0.885003370092833\,31$	$a_{54} = -1.0620040441114\,01$	$c_{54} = 14.160053921485\,337$
$\tilde{a}_{61} = -1.7644376487744849$	$\gamma_{61} = 1.684069277985373\,3$	$a_{61} = -7.1704549624232\,04$	$c_{61} = 34.626058309305\,947$
$\tilde{a}_{62} = -0.4747565572063031\,7$	$\gamma_{62} = 0.4182659436138561\,4$	$a_{62} = -4.7416366714818\,72$	$c_{62} = 15.300849761145\,031$
$\tilde{a}_{63} = 2.369691846915804\,8$	$\gamma_{63} = -1.881406216873008$	$a_{63} = -16.310026313309\,713$	$c_{63} = 56.995557866267\,57$
$\tilde{a}_{64} = 0.6195025359064983\,32$	$\gamma_{64} = -0.1137861475833642\,8$	$a_{64} = -1.0620040441114\,01$	$c_{64} = 18.408070097930\,938$
$\tilde{a}_{65} = 0.25$	$\gamma_{65} = -0.3571428571428571\,4$	$a_{65} = 1.0$	$c_{65} = -5.7142857142857\,14$
$b_1 = -0.0803683707089111\,65$	$\hat{b}_1 = -1.764437648774484\,9$	$m_1 = -7.1704549624232\,198$	$\hat{m}_1 = -7.1704549624232\,32$
$b_2 = -0.0564906135924470\,36$	$\hat{b}_2 = -0.4747565572063031\,7$	$m_2 = -4.7416366714818\,74$	$\hat{m}_2 = -4.7416366714818\,75$
$b_3 = 0.4882856300427967\,9$	$\hat{b}_3 = 2.369691846915804\,8$	$m_3 = -16.310026313309\,713$	$\hat{m}_3 = -16.310026313309\,713$
$b_4 = 0.5057162114816190\,4$	$\hat{b}_4 = 0.6195025359064983\,32$	$m_4 = -1.0620040441114\,01$	$\hat{m}_4 = -1.0620040441114\,01$
$b_5 = -0.1071428571428571\,4$	$\hat{b}_5 = 0.25$	$m_5 = 1.0$	$\hat{m}_5 = 1.0$
$b_6 = 0.25$	$\hat{b}_6 = 0.0$	$m_6 = 1.0$	$\hat{m}_6 = 0.0$

TABLE B.14: Coefficients for RODASP

0	0	0	0	0
	12/23	0	0	0
	-68/375	368/375	0	0
	31/144	529/1152	125/384	0

TABLE B.15: Butcher array of the four-stage CERK method

k	b_{k1}	b_{k2}	b_{k3}	b_{k4}
0	1	0	0	0
1	-65/48	529/384	125/128	-1
2	41/72	-529/576	-125/192	1

TABLE B.16: Coefficients of the continuous extension of the four-stage CERK method

Bibliography

[1] P. Anninos, C. Bryant, P. C. Fragile, A. M. Holgado, C. Lau, and D. Nemergut. CosmosDG: An hp-adaptive discontinuous Galerkin code for hyper-resolved relativistic MHD. *Astrophys J., Suppl. Ser.*, 231(2):17, 2017.

[2] D. Appelö and T. Hagstrom. A general perfectly matched layer model for hyperbolic-parabolic systems. *SIAM J. Sci. Comput.*, 13(5):3301–3323, 2009.

[3] Argonne National Laboratory. MPICH, http://www.mcs.anl.gov/research/projects/mpich2/, last accessed on 4/12/2020.

[4] D. N. Arnold, F. Brezzi, B. Cockburn, and L. D. Marini. Unified analysis of discontinuous Galerkin methods for elliptic problems. *SIAM J. Num. Anal.*, 39(5):1749–1779, 2002.

[5] I. Babuska, B. A. Szabo, and I. N. Katz. The p-version of the finite element method. *SIAM J. Numer. Anal.*, 18(3):515–545, 1981.

[6] K. J. Badcock, I. C. Glover, and B. E. Richards. A preconditioner for steady two-dimensional turbulent flow simulation. *Int. J. Num. Meth. Heat Fluid Flow*, 6:79–93, 1996.

[7] B. S. Baldwin and H. Lomax. Thin layer approximation and algebraic model for separated turbulent flows. *AIAA Paper 78-257*, 1978.

[8] D. S. Bale, R. J. LeVeque, S. Mitran, and J. A. Rossmanith. A wave propagation method for conservation laws and balance laws with spatially varying flux functions. *SIAM J. Sci. Comput.*, 24(3):955–978, 2002.

[9] A. L. Banka. Practical Applications of CFD in heat processing. *Heat Treating Progress*, August:44–49, 2005.

[10] W. Barsukow, P. V. F. Edelmann, C. Klingenberg, F. Miczek, and F. K. Röpke. A numerical scheme for the compressible low-Mach number regime of ideal fluid dynamics. *J. Sci. Comput.*, 72(2):623–646, 2017.

[11] T. Barth and M. Ohlberger. Finite volume methods: foundation and analysis. In E. Stein, R. de Borst, and T. J. R. Hughes, editors, *Encyclopedia of Computational Mechanics*, volume 1: Fundame, chapter 15, pages 439–473. John Wiley and Sons, 2004.

[12] T. J. Barth and D. C. Jesperson. The design and application of upwind schemes on unstructured meshes. *AIAA Paper 89-0366*, 1989.

[13] F. Bassi, A. Ghidoni, and S. Rebay. Optimal Runge-Kutta smoothers for the p-multigrid discontinuous Galerkin solution of the 1D Euler equations. *J. Comp. Phys.*, 11:4153–4175, 2011.

[14] F. Bassi, A. Ghidoni, S. Rebay, and P. Tesini. High-order accurate p-multigrid discontinuous Galerkin solution of the Euler equations. *Int. J. Num. Meth. Fluids*, 60:847–865, 2009.

[15] F. Bassi and S. Rebay. Numerical evaluation of two discontinuous Galerkin methods for the compressible Navier-Stokes equations. *Int. J. Num. Fluids*, 40:197–207, 2002.

[16] R. Becker and R. Rannacher. An optimal control approach to a posteriori error estimation in finite element methods. *Acta Numerica*, 10:1–102, 2001.

[17] O. O. Bendiksen. A new approach to computational aeroelasticity. *AIAA Paper AIAA-91-0939-CP*, 1991.

[18] M. Benzi and D. Bertaccini. Approximate inverse preconditioning for shifted linear systems. *BIT*, 43:231–244, 2003.

[19] M. Benzi and M. Tuma. Orderings for factorized sparse approximate inverse preconditioners. *SIAM J. Sci. Comput.*, 21:1851–1868, 2000.

[20] M. J. Berger and P. Colella. Local Adaptive Mesh Refinement for Shock Hydrodynamics. *J. Comp. Phys.*, 82:64–84, 1989.

[21] D. Bertaccini, M. Donatelli, F. Durastante, and S. Serra-Capizzano. Optimizing a multigrid Runge-Kutta smoother for variable-coefficient convection-diffusion equations. *Lin. Alg. Appl.*, 533:507–535, 2017.

[22] H. Bijl and M. H. Carpenter. Iterative solution techniques for unsteady flow computations using higher order time integration schemes. *Int. J. Num. Meth. in Fluids*, 47:857–862, 2005.

[23] H. Bijl, M. H. Carpenter, V. N. Vatsa, and C. A. Kennedy. Implicit time integration schemes for the unsteady compressible Navier-Stokes equations: laminar flow. *J. Comp. Phys.*, 179:313–329, 2002.

[24] P. Birken. Numerical simulation of tunnel fires using preconditioned finite volume schemes. *ZAMP*, 59:416–433, 2008.

[25] P. Birken. *Numerical Methods for the Unsteady Compressible Navier-Stokes Equations*. Habilitation Thesis. University of Kassel, 2012.

[26] P. Birken. Optimizing Runge-Kutta smoothers for unsteady flow problems. *ETNA*, 39:298–312, 2012.

[27] P. Birken. Termination criteria for inexact fixed point schemes. *Numer. Linear Algebra Appl.*, 22(4):702–716, 2015.

[28] P. Birken, J. Bull, and A. Jameson. A note on terminology in multigrid methods. In *PAMM*, volume 16, pages 721–722, 2016.

[29] P. Birken, J. Bull, and A. Jameson. A study of multigrid smoothers used in compressible CFD based on the convection diffusion equation. In M. Papadrakakis, V. Papadopoulos, G. Stefanou, and V. Plevris, editors, *ECCOMAS Congress 2016, VII European Congress on Computational Methods in Applied Sciences and Engineering*, volume 2, pages 2648–2663, Crete Island, Greece, 2016.

[30] P. Birken, J. Bull, and A. Jameson. Preconditioned smoothers for the full approximation scheme for the RANS equations. *Journal of Scientific Computing*, 78(2):995–1022, 2019.

[31] P. Birken, J. Duintjer Tebbens, A. Meister, and M. Tuma. Updating preconditioners for permuted non-symmetric linear systems. *PAMM*, 7(1):1022101–1022102, 2007.

[32] P. Birken, J. Duintjer Tebbens, A. Meister, and M. Tuma. Preconditioner updates applied to CFD model problems. *Appl. Num. Math.*, 58:1628–1641, 2008.

[33] P. Birken, G. Gassner, M. Haas, and C.-D. Munz. Efficient time integration for discontinuous Galerkin methods for the unsteady 3D Navier-Stokes equations. In J. Eberhardsteiner, editor, *European Congress on Computational Methods and Applied Sciences and Engineering (ECCOMAS 2012)*, 2012.

[34] P. Birken, G. Gassner, M. Haas, and C.-D. Munz. Preconditioning for modal discontinuous Galerkin methods for unsteady 3D Navier-Stokes equations. *J. Comp. Phys.*, 240:20–35, 2013.

[35] P. Birken, G. J. Gassner, and L. M. Versbach. Subcell finite volume multigrid preconditioning for high-order discontinuous Galerkin methods. *Int. J. CFD*, 33(9):353–361, 2019.

[36] P. Birken, T. Gleim, D. Kuhl, and A. Meister. Fast solvers for unsteady thermal fluid structure interaction. *Int. J. Num. Meth. Fluids*, 79(1):16–29, 2015.

[37] P. Birken and A. Jameson. Nonlinear iterative solvers for unsteady Navier-Stokes equations. In *Proceedings of Hyp2008—the Twelfth International Conference on Hyperbolic Problems*, pages 429–438. AMS, 2009.

[38] P. Birken and A. Jameson. On nonlinear preconditioners in Newton-Krylov-Methods for unsteady flows. *Int. J. Num. Meth. Fluids*, 62:565–573, 2010.

[39] P. Birken and A. Meister. Stability of preconditioned finite volume schemes at low Mach numbers. *BIT*, 45(3):463–480, 2005.

[40] P. Birken and A. Monge. Numerical methods for unsteady thermal fluid structure interaction. In S. Frei, B. Holm, T. Richter, T. Wick, and H. Yang, editors, *Fluid-Structure Interaction*, volume 20 of *Radon Series on Computational and Applied Mathematics*, chapter 4, pages 129–168. de Gruyter, 2018.

[41] P. Birken, K. J. Quint, S. Hartmann, and A. Meister. A time-adaptive fluid-structure interaction method for thermal coupling. *Comp. Vis. in Science*, 13(7):331–340, 2010.

[42] C. Bischof and A. Carle. ADIFOR, http://www.mcs.anl.gov/research/projects/adifor/, last accessed 4/12/2020.

[43] S. Blanes and F. Casas. On the necessity of negative coefficients for operator splitting schemes of order higher than two. *Appl. Num. Math.*, 54(1):23–37, 2005.

[44] J. Blazek. *Computational Fluid Dynamics*. Elsevier, 2nd edition, 2004.

[45] D. S. Blom, P. Birken, H. Bijl, F. Kessels, A. Meister, and A. H. van Zuijlen. A comparison of Rosenbrock and ESDIRK methods combined with iterative solvers for unsteady compressible flows. *Adv. Comput. Math.*, 42(6):1401–1426, 2016.

[46] M. Bolten, D. Moser, and R. Speck. Asymptotic convergence of the parallel full approximation scheme in space and time for linear problems. *Num. Lin. Algebra with Appl.*, pages 1–24, 2017.

[47] S. Brdar, A. Dedner, and R. Klöfkorn. Compact and stable discontinuous Galerkin methods for convection-diffusion problems. *SIAM J. Sci. Comput.*, 34(1):A263–A282, 2012.

[48] K. E. Brenan, S. L. Campbell, and L. R. Petzold. *Numerical Solution of Initial-Value Problems in Differential Algebraic Equations*, volume 14. SIAM, Classics in Applied Mathematics, 1996.

[49] P. Bruel, S. Delmas, J. Jung, and V. Perrier. A low Mach correction able to deal with low Mach acoustics. *J. Comp. Phys.*, 378:723–759, 2019.

[50] P. R. Brune, M. G. Knepley, B. F. Smith, and X. Tu. Composing scalable nonlinear algebraic solvers. *SIAM Rev.*, 57(4):535–565, 2015.

[51] J. M. Buchlin. Convective heat transfer and infrared thermography. *J. Appl. Fluid Mech.*, 3(1):55–62, 2010.

[52] J. C. Butcher. Linear and non-linear stability for general linear methods. *BIT Numerical Mathematics*, 27(2):181–189, 1987.

[53] M. H. Carpenter, D. Gottlieb, and S. Abarbanel. The stability of numerical boundary treatments for compact high-order finite-difference schemes. *J. Comput. Phys.*, 108:272–295, 1993.

[54] M. H. Carpenter and C. A. Kennedy. Fourth-order 2N-storage Runge-Kutta schemes. Technical report, NASA, 1994.

[55] D. A. Caughey and A. Jameson. How many steps are required to solve the Euler equations of steady compressible flow: in search of a fast solution algorithm. *AIAA Paper 2001-2673*, 2001.

[56] CENTAUR Software. CENTAUR Grid Generator, http://www.centaursoft.com/grid-generator, last accessed on 4/12/2020.

[57] G.-Q. Chen and D. Wang. The Cauchy problem for the Euler equations for compressible fluids. In *Handbok on Mathematical Fluid Dynamics*, volume 1. Elsevier, 2002.

[58] E. Chiodaroli, C. De Lellis, and O. Kreml. Global ill-posedness of the isentropic system of gas dynamics. *Communications on Pure and Applied Mathematics*, 68(7):1157–1190, 2015.

[59] A. J. Chorin and J. Marsden. *A Mathematical Introduction to Fluid Mechanics*. Springer Verlag, New York, 1997.

[60] A. Christlieb, B. Ong, and J.-M. Qiu. Integral deferred correction methods constructed with high order Runge-Kutta integrators. *Math. Comput.*, 79(270):761–783, 2010.

[61] B. Cockburn and C.-W. Shu. The local discontinuous Galerkin method for time-dependent convection diffusion systems. *SIAM J. Num. Analysis*, 35:2440–2463, 1998.

[62] B. Cockburn and C.-W. Shu. Runge–Kutta discontinuous Galerkin methods for convection-dominated problems. *J. Sci. Comput.*, 16(3):173–261, 2001.

[63] T. Colonius and S. K. Lele. Computational aeroacoustics: progress on nonlinear problems of sound generation. *Progress in Aerospace Sciences*, 40(6):345–416, 2004.

[64] G. J. Cooper and A. Sayfy. Additive methods for the numerical solution of ordinary differential equations. *Math. Comput.*, 35(152):1159–1172, 1980.

[65] R. Courant, K. O. Friedrichs, and H. Lewy. Über die partiellen differentialgleichungen der mathematischen Physik. *Math. Annalen*, 100:32–74, 1928.

[66] M. Crandall and A. Majda. The method of fractional steps for conservation laws. *Numer. Math.*, 34:285–314, 1980.

[67] G. A. Davis and O. O. Bendiksen. Transonic panel flutter. *AIAA Paper 93-1476*, 1993.

[68] F. Davoudzadeh, H. Mcdonald, and B. E. Thompson. Accuracy evaluation of unsteady CFD numerical schemes by vortex preservation. *Comput. Fluids*, 24:883–895, 1995.

[69] B. de St. Venant. Memoire sur la dynamique des fluides. *C. R. Acad. Sci. Paris*, 17:1240–1242, 1845.

[70] A. Dedner, C. Makridakis, and M. Ohlberger. Error control for a class of Runge–Kutta discontinuous Galerkin methods for Nonlinear Conservation Laws. *SIAM J. Numerical Analysis*, 45(2):514, 2007.

[71] A. Dedner and M. Ohlberger. A new hp-adaptive DG scheme for conservation laws based on error control. In S. Benzoni-Gavage and D. Serre, editors, *Hyperbolic Problems: Theory, Numerics, Applications, ISNM, Vol. 141*. Springer, Berlin, Heidelberg, 2008.

[72] K. Dekker and J. G. Verwer. *Stability of Runge-Kutta Methods for Stiff Nonlinear Differential Equations*. CWI Monogr. 2. North Holland, Amsterdam, 1984.

[73] S. Dellacherie. Analysis of Godunov type schemes applied to the compressible Euler system at low Mach number. *J. Comp. Phys.*, 229(4):978–1016, 2010.

[74] R. Dembo, R. Eisenstat, and T. Steihaug. Inexact Newton methods. *SIAM J. Numer. Anal.*, 19:400–408, 1982.

[75] J. E. Dennis and R. B. Schnabel. *Numerical Methods for Unconstrained Optimization and Nonlinear Equations*. Classics in Applied Mathematics. SIAM, Philadelphia, 1996.

[76] P. Deuflhard. *Newton Methods*. Springer, 2004.

[77] M. Donatelli, C. Garoni, C. Manni, and H. Speleers. Symbol-based multigrid methods for Galerkin B-spline isogeometric analysis. *SIAM J. Num. Anal.*, 55(1):31–62, 2017.

[78] J. Duintjer Tebbens and M. Tuma. Efficient preconditioning of sequences of nonsymmetric linear systems. *SIAM J. Sci. Comput.*, 29(5):1918–1941, 2007.

[79] J. Duintjer Tebbens and M. Tuma. Preconditioner updates for solving sequences of linear systems in matrix-free environment. *Num. Lin. Algebra with Appl.*, 17:997–1019, 2010.

[80] A. Dutt, L. Greengard, and V. Rokhlin. Spectral deferred correction methods for ordinary differential equations. *BIT Numerical Mathematics*, 40(2):241–266, 2000.

[81] R. P. Dwight. *Efficiency Improvements of RANS-Based Analysis and Optimization Using Implicit and Adjoint Methods on Unstructured Grids*. Phd thesis, University of Manchester, 2006.

[82] S. C. Eisenstat, H. C. Elman, and M. H. Schultz. Variational iterative methods for nonsymmetric systems of linear equations. *SIAM J. Num. Anal.*, 20(2):345–357, 1983.

[83] S. C. Eisenstat and H. F. Walker. Choosing the forcing terms in an inexact Newton method. *SIAM J. Sci. Comput.*, 17(1):16–32, 1996.

[84] P. Ellsiepen. *Zeits- und ortsadaptive Verfahren angewandt auf Mehrphasenprobleme poröser Medien*. Dissertation, University of Stuttgart, Institute of Mechanics II, 1999.

[85] M. Emmett and M. Minion. Toward an efficient parallel in time method for partial differential equations. *Comm. App. Math. and Comp. Sci.*, 7(1):105–132, 2012.

[86] F. F. F. Grinstein, L. G. Len G. Margolin, and W. J. Rider, editors. *Implicit Large Eddy Simulation—Computing Turbulent Fluid Dynamics*. Oxford University Press, 2011.

[87] V. Faber and T. Manteuffel. Necessary and sufficient conditions for the existence of a conjugate gradient method. *SIAM J. Numer. Anal.*, 21(2):352–362, 1984.

[88] R. D. Falgout, S. Friedhoff, Tz. V. Kolev, S. P. MacLachlan, and J. B. Schroder. Parallel time integration with multigrid. *SIAM J. Sci. Comput.*, 36(6):C635–C661, 2014.

[89] C. Farhat. CFD-based nonlinear computational aeroelasticity. In E. Stein, R. de Borst, and T. J. R. Hughes, editors, *Encyclopedia of Computational Mechanics*, volume 3: Fluids, chapter 13, pages 459–480. John Wiley & Sons, 2004.

[90] K. J. Fidkowski, T. A. Oliver, J. Lu, and D. L. Darmofal. p-Multigrid solution of high-order discontinuous Galerkin discretizations of the compressible Navier-Stokes equations. *J. Comp. Phys.*, 207:92–113, 2005.

[91] K. J. Fidkowski and P. L. Roe. An entropy adjoint approach to mesh refinement. *SIAM J. Sci. Comput.*, 32(3):1261–1287, 2010.

[92] U. S. Fjordholm, S. Mishra, and E. Tadmor. On the computation of measure-valued solutions. *Acta Numerica*, 25:567–679, 2016.

[93] L. Friedrich, G. Schnücke, A. R. Winters, D. C. Del Rey Fernández, G. J. Gassner, and M. H. Carpenter. Entropy stable space–time discontinuous Galerkin schemes with summation-by-parts property for hyperbolic conservation laws. *J. Sci. Comput.*, 80(1):175–222, 2019.

[94] M. J. Gander. 50 years of time parallel time integration. In T. Carraro, M. Geiger, S. Körkel, and R. Rannacher, editors, *Multiple Shooting and Time Domain Decomposition Methods*, volume 9 of *Contributions in Mathematical and Computational Sciences*, pages 183–202. Springer International Publishing, 2015.

[95] M. J. Gander, F. Kwok, and B. C. Mandal. Dirichlet-Neumann and Neumann-Neumann waveform relaxation algorithms for parabolic problems. *ETNA*, 45:424–456, 2016.

[96] M. J. Gander and S. Vandewalle. Analysis of the parareal time-parallel time-integration method. *SIAM J. Sci. Comput.*, 29(2):556–578, 2007.

[97] C. Garoni and S. Serra-Capizzano. *Generalized Locally Toeplitz Sequences: Theory and Applications*, volume 1. Springer, 2018.

[98] G. Gassner, M. Dumbser, F. Hindenlang, and C.-D. Munz. Explicit one-step time discretizations for discontinuous Galerkin and finite volume schemes based on local predictors. *J. Comp. Phys.*, 230(11):4232–4247, 2011.

[99] G. Gassner, F. Lörcher, and C.-D. Munz. A contribution to the construction of diffusion fluxes for finite volume and discontinuous Galerkin schemes. *J. Comp. Phys.*, 224:1049–1063, 2007.

[100] G. Gassner, F. Lörcher, and C.-D. Munz. A discontinuous Galerkin scheme based on a space-time expansion II. Viscous flow equations in multi dimensions. *J. Sci. Comput.*, 34:260–286, 2008.

[101] G. J. Gassner. A skew-symmetric discontinuous galerkin spectral element discretization and its relation to sbp-sat finite difference methods. *SIAM J. Sci. Comput.*, 35(3):1233–1253, 2013.

[102] G. J. Gassner, F. Lörcher, C.-D. Munz, and J. S. Hesthaven. Polymorphic nodal elements and their application in discontinuous Galerkin methods. *J. Comp. Phys.*, 228(5):1573–1590, 2009.

[103] S. Gaudreault, G. Rainwater, and M. Tokman. KIOPS: A fast adaptive Krylov subspace solver for exponential integrators. *J. Comp. Phys.*, 372:236–255, 2018.

[104] T. Gerhold, O. Friedrich, J. Evans, and M. Galle. Calculation of complex three-dimensional configurations employing the DLR-TAU-Code. *AIAA Paper 97-0167*, January, 1997.

[105] C. Geuzaine and J.-F. Remacle. Gmsh, http://www.geuz.org/gmsh/, last accessed on 4/12/2020.

[106] M. B. Giles. Stability analysis of numerical interface conditions in fluid-structure thermal analysis. *Int. J. Num. Meth. Fluids*, 25:421–436, 1997.

[107] E. Godlewski and P.-A. Raviart. *Numerical Approximation of Hyperbolic Systems of Conservation Laws*, volume 118 of *Applied Mathematical Sciences*. Springer, New York, Berlin, Heidelberg, 1996.

[108] S. K. Godunov. A finite difference method for the numerical computation of discontinuous solutions of the equations of fluid dynamics. *Mat. Sb.*, 47:357–393, 1959.

[109] J. B. Goodman and R. J. LeVeque. On the accuracy of stable schemes for 2D scalar conservation laws. *Math. Comp.*, 45(171):15–21, 1985.

[110] M. Görtz and P. Birken. On the convergence rate of the Dirichlet-Neumann iteration for coupled Poisson problems on unstructured grids. In R. Klöfkorn, E. Keilegavlen, F. A. Radu, and J. Fuhrmann, editors, *Finite Volumes for Complex Applications IX—Methods, Theoretical Aspects, Examples*, volume 323 of *Springer Proceedings in Mathematics & Statistics*, pages 355–363. Springer International Publishing, 2020.

[111] S. Gottlieb, D. I. Ketcheson, and C.-W. Shu. High order strong stability preserving time discretizations. *J. Sci. Comp.*, 38:251–289, 2009.

[112] S. Gottlieb, C.-W. Shu, and E. Tadmor. Strong stability-preserving high-order time discretization methods. *SIAM Rev.*, 43(1):89–112, 2001.

[113] A. Greenbaum, V. Pták, and Z. Strakoš. Any nonincreasing convergence curve is possible for GMRES. *SIAM J. Matrix Anal. Appl.*, 17(3):465–469, 1996.

[114] J. J. Greenberg and A. Y. Leroux. A well-balanced scheme for the numerical processing of source terms in hyperbolic equations. *SIAM J. Numer. Anal.*, 33(1):1–16, 1996.

[115] M. J. Grote and T. Huckle. Parallel preconditioning with sparse approximate inverses. *SIAM J. Sci. Comput.*, 18(3):838–853, 1997.

[116] H. Guillard and C. Farhat. On the significance of the geometric conservation law for flow computations on moving meshes. *Comp. Meth. Appl. Mech. Engrg.*, 190:1467–1482, 2000.

[117] H. Guillard and A. Murrone. On the behavior of upwind schemes in the low Mach number limit: II. Godunov type schemes. *Comput. Fluids*, 33:655–675, may 2004.

[118] H. Guillard and C. Viozat. On the behaviour of upwind schemes in the low Mach number limit. *Comput. Fluids*, 28:63–86, 1999.

[119] B. Gustafsson. *High Order Difference Methods for Time Dependent PDE*. Springer, Berlin, Heidelberg, 2008.

[120] W. Hackbusch. *Multi-Grid Methods and Applications*, volume 4 of *Springer Series in Computational Mathematics*. Springer, Berlin, Heidelberg, New York, Tokio, 1985.

[121] E. Hairer, S. P. Nørsett, and G. Wanner. *Solving Ordinary Differential Equations I*. Springer, Berlin, Heidelberg, New York, 2000.

[122] E. Hairer and G. Wanner. *Solving Ordinary Differential Equations II*. Springer, Berlin, second rev. edition, 2002.

[123] A. Harten, J. M. Hyman, and P. D. Lax. On finite-difference approximations and entropy conditions for shocks. *Comm. Pure Appl. Math.*, XXIX:297–322, 1976.

[124] R. Hartmann. Adaptive discontinuous Galerkin methods with shock-capturing for the compressible Navier-Stokes equations. *Int. J. Num. Meth. Fluids*, 51:1131–1156, 2006.

[125] S. Hartmann. *Finite-Elemente Berechnung Inelastischer Kontinua*. PhD thesis, Kassel, 2003.

[126] S. Hartmann, J. Duintjer Tebbens, K. J. Quint, and A. Meister. Iterative solvers within sequences of large linear systems in non-linear structural mechanics. *ZAMM*, 89(9):711–728, 2009.

[127] L. Hascoët and V. Pascual. The tapenade automatic differentiation tool: principles, model, and specification. *ACM TOMS*, 39(3), 2013.

[128] P. Haupt. *Continuum Mechanics and Theory of Materials*. Springer, Berlin, 2000.

[129] U. Heck, U. Fritsching, and K. Bauckhage. Fluid flow and heat transfer in gas jet quenching of a cylinder. *International Journal of Numerical Methods for Heat & Fluid Flow*, 11:36–49, 2001.

[130] W. D. Henshaw and K. K. Chand. A composite grid solver for conjugate heat transfer in fluid–structure systems. *J. Comp. Phys.*, 228(10):3708–3741, 2009.

[131] J. S. Hesthaven and T. Warburton. *Nodal Discontinuous Galerkin Methods: Algorithms, Analysis, and Applications*. Springer, 2008.

[132] F. Hindenlang, T. Bolemann, and C.-D. Munz. Mesh curving techniques for high order discontinuous Galerkin simulations. In N. Kroll, C. Hirsch, F. Bassi, C. Johnston, and K. Hillewaert, editors, *IDIHOM: Industrialization of High-Order Methods-A Top-Down Approach*, pages 133–152. Springer, 2015.

[133] M. Hinderks and R. Radespiel. Investigation of hypersonic gap flow of a reentry nosecap with consideration of fluid structure interaction. *AIAA Paper, 06-1111*, 2006.

[134] A. C. Hindmarsh, P. N. Brown, K. E. Grant, S. L. Lee, R. Serban, D. E. Shumaker, and C. S. Woodward. SUNDIALS: Suite of nonlinear and differential/algebraic equation solvers. *ACM TOMS*, 31(3):363–396, 2005.

[135] C. Hirsch. *Numerical Computation of Internal and External Flows*, volume 1. Wiley & Sons, Chichester, New York, 1988.

[136] C. Hirsch. *Numerical Computation of Internal and External Flows*, volume 2. Wiley & Sons, Chichester, New York, 1988.

[137] M. Hochbruck and A. Ostermann. Exponential integrators. *Acta Numerica*, 19:209–286, 2010.

[138] W. Hundsdorfer and J. G. Verwer. *Numerical Solution of Time-Dependent Advection-Diffusion-Reaction Equations*, volume 33 of *Series in Computational Mathematics*. Springer, New York, Berlin, Heidelberg, 2003.

[139] H. T. Huynh. A flux reconstruction approach to high-order schemes including discontinuous Galerkin methods. *AIAA Paper AIAA 2007-4079*, June, 2007.

[140] A. Jameson. Evaluation of fully implicit Runge Kutta schemes for unsteady flow calculations. *J. Sci. Comput.*, 73:819–852, 2017.

[141] A. Jameson. Transonic flow calculations for aircraft. In F. Brezzi, editor, *Numerical Methods in Fluid Dynamics*, Lecture Notes in Mathematics, pages 156–242. Springer, 1985.

[142] A. Jameson. Time dependent calculations using multigrid, with applications to unsteady flows past airfoils and wings. *AIAA Paper 91-1596*, June, 1991.

[143] A. Jameson. Aerodynamics. In E. Stein, R. de Borst, and T. J. R. Hughes, editors, *Encyclopedia of Computational Mechanics*, volume 3: Fluids, chapter 11, pages 325–406. John Wiley & Sons, 2004.

[144] A. Jameson. Application of dual time stepping to fully implicit Runge Kutta schemes for unsteady flow calculations. *AIAA Paper 2015-2753*, 2015.

[145] A. Jameson and T. J. Baker. Solution of the Euler equations for complex configurations. *AIAA Paper 83-1929*, pages 293–302, 1983.

[146] A. Jameson, W. Schmidt, and E. Turkel. Numerical solution of the Euler equations by finite volume methods using Runge-Kutta time-stepping schemes. *AIAA Paper 1981-1259*, June, 1981.

[147] J. Jeong and F. Hussain. On the identification of a vortex. *J. Fluid Mech.*, 285:69–94, 1995.

[148] V. John and J. Rang. Adaptive time step control for the incompressible Navier-Stokes equations. *Comp. Meth. Appl. Mech. Engrg.*, 199:514–524, 2010.

[149] G. Jothiprasad, D. J. Mavriplis, and D. A. Caughey. Higher-order time integration schemes for the unsteady Navier-Stokes equations on unstructured meshes. *J. Comp. Phys.*, 191:542–566, 2003.

[150] A. Kanevsky, M. H. Carpenter, D. Gottlieb, and J. S. Hesthaven. Application of implicit-explicit high order Runge-Kutta methods to discontinuous-Galerkin schemes. *J. Comp. Phys.*, 225:1753–1781, 2007.

[151] R. Kannan and Z. J. Wang. A study of viscous flux formulations for a p-multigrid spectral volume Navier Stokes solver. *J. Sci. Comput.*, 41(2):165–199, 2009.

[152] G. E. Karniadakis and S. Sherwin. *Spectral/hp Element Methods for Computational Fluid Dynamics*. Oxford University Press, 2nd edition, 2005.

[153] Karypis Lab. ParMETIS, http://glaros.dtc.umn.edu/gkhome/metis/parmetis/overview, last accessed on 4/12/20.

[154] C. T. Kelley. *Iterative Methods for Linear and Nonlinear Equations*. SIAM, Philadelphia, PA, 1995.

[155] C. A. Kennedy and M. H. Carpenter. Additive Runge-Kutta schemes for convection-diffusion-reaction equations. *Appl. Num. Math.*, 44:139–181, 2003.

[156] D. I. Ketcheson, C. B. Macdonald, and S. Gottlieb. Optimal implicit strong stability preserving Runge-Kutta methods. *Appl. Num. Math.*, 59:373–392, 2009.

[157] C. M. Klaij, J. J. W. van der Vegt, and H. van der Ven. Space-time discontinuous Galerkin method for the compressible Navier-Stokes equations. *J. Comp. Phys.*, 217(2):589–611, 2006.

[158] C. M. Klaij, M. H. van Raalte, J. J. W. van der Vegt, and H. van der Ven. h-Multigrid for space-time discontinuous Galerkin discretizations of the compressible Navier-Stokes equations. *J. Comp. Phys.*, 227:1024–1045, 2007.

[159] H. Klimach, J. Zudrop, and S. Roller. Generation of high order geometry representations in octree meshes. Technical report, University of Siegen, 2015.

[160] D. A. Knoll and D. E. Keyes. Jacobian-free Newton-Krylov methods: a survey of approaches and applications. *J. Comp. Phys.*, 193:357–397, 2004.

[161] D. A. Kopriva. *Implementing Spectral Methods for Partial Differential Equations*. Springer, 2000.

[162] D. A. Kopriva, S. L. Woodruff, and M. Y. Hussaini. Computation of electromagnetic scattering with a non-conforming discontinuous spectral element method. *Int. J. Num. Meth. in Eng.*, 53:105–122, 2002.

[163] H.-O. Kreiss and J. Lorenz. *Initial Boundary Value Problems and the Navier-Stokes Equations*. Academic Press, New York, 1989.

[164] D. Kröner and M. Ohlberger. A posteriori error estimates for upwind finite volume schemes for nonlinear conservation laws in multi dimensions. *Math. Comp.*, 69(229):25–39, 1999.

[165] U. Küttler and W. A. Wall. Fixed-point fluid–structure interaction solvers with dynamic relaxation. *Comput. Mech.*, 43:61–72, 2008.

[166] S. Langer. Investigation and application of point implicit Runge-Kutta methods to inviscid flow problems. *Int. J. Num. Meth. Fluids*, 69(2):332–352, 2011.

[167] S. Langer. Application of a line implicit method to fully coupled system of equations for turbulent flow problems. *Int. J. CFD*, 27(3):131–150, 2013.

[168] S. Langer. Agglomeration multigrid methods with implicit Runge–Kutta smoothers applied to aerodynamic simulations on unstructured grids. *J. Comp. Phys.*, 277:72–100, 2014.

[169] S. Langer and D. Li. Application of point implicit Runge-Kutta methods to inviscid and laminar flow problems using AUSM and AUSM + upwinding. *International Journal of Computational Fluid Dynamics*, 25(5):255–269, 2011.

[170] S. Langer, A. Schwöppe, and N. Kroll. Investigation and comparison of implicit smoothers applied in agglomeration multigrid. *AIAA Journal*, 53(8):2080–2096, 2015.

[171] J. O. Langseth, A. Tveito, and R. Winther. On the convergence of operator splitting applied to conversation laws with source terms. *SIAM J. Num. Anal.*, 33(3):843–863, 1996.

[172] P. Lax and B. Wendroff. Systems of conservation laws. *Comm. Pure Appl. Math.*, 13:217–237, 1960.

[173] P. Le Tallec and J. Mouro. Fluid structure interaction with large structural displacements. *Comp. Meth. Appl. Mech. Engrg.*, 190:3039–3067, 2001.

[174] R. J. LeVeque. *Numerical Methods for Conservation Laws*. Birkhäuser, Basel, 1990.

[175] R. J. LeVeque. *Finite Volume methods for Hyperbolic Problems*. Cambridge University Press, Cambridge, 2002.

[176] P. T. Lin, J. N. Shadid, and P. H. Tsuji. On the performance of Krylov smoothing for fully coupled AMG preconditioners for VMS resistive MHD. *Int. J. Num. Meth. Eng.*, 120(12):1297–1309, 2019.

[177] P.-L. Lions. *Mathematical Topics in Fluid Mechanics Volume 2 Compressible Models*. Oxford Science Publications, 1998.

[178] N. Lior. The cooling process in gas quenching. *J. Materials Processing Technology*, 155-156:1881–1888, 2004.

[179] M.-S. Liou. A sequel to AUSM, Part II: AUSM+-up for all speeds. *J. Comp. Phys.*, 214:137–170, 2006.

[180] F. Lörcher, G. Gassner, and C.-D. Munz. A discontinuous Galerkin scheme based on a space-time expansion. I. Inviscid compressible flow in one space dimension. *Journal of Scientific Computing*, 32(2):175–199, 2007.

[181] Ch. Lubich and A. Ostermann Linearly implicit time discretization of non-linear parabolic equations. *IMA J. Num. Analysis*, 15(4): 555–583, 1995.

[182] H. Luo, J. D. Baum, and R. Löhner. A p-multigrid discontinuous Galerkin method for the Euler equations on unstructured grids. *J. Comp. Phys.*, 211:767–783, 2006.

[183] K. T. Mandli, A. J. Ahmadia, M. Berger, D. Calhoun, D. L. George, Y. Hadjimichael, D. I. Ketcheson, G. I. Lemoine, and R. J. LeVeque. Clawpack: Building an open source ecosystem for solving hyperbolic PDEs. *PeerJ Computer Science*, 2016(8):1–27, 2016.

[184] K. Mani and D. J. Mavriplis. Efficient solutions of the Euler equations in a time-adaptive space-time framework. *AIAA-Paper 2011-774*, January, 2011.

[185] M. F. Maritz and S. W. Schoombie. Exact analysis of nonlinear instability in a discrete Burgers' equation. *J. Comp. Phys.*, 97(1):73–90, 1991.

[186] R. Massjung. *Numerical Schemes and Well-Posedness in Nonlinear Aeroelasticity*. PhD thesis, RWTH Aachen, 2002.

[187] R. Massjung. Discrete conservation and coupling strategies in nonlinear aeroelasticity. *Comp. Meth. Appl. Mech. Engrg.*, 196:91–102, 2006.

[188] H. G. Matthies, R. Niekamp, and J. Steindorf. Algorithms for strong coupling procedures. *Comput. Methods Appl. Mech. Engrg.*, 195:2028–2049, 2006.

[189] H. G. Matthies and J. Steindorf. Partitioned strong coupling algorithms for fluid-structure interaction. *Comput. Struct.*, 81(8-11):805–812, may 2003.

[190] D. J. Mavriplis. Directional agglomeration multigrid techniques for high-Reynolds number viscous flows. Technical Report 98-7, ICASE, 1998.

[191] D. J. Mavriplis. An assessment of linear versus nonlinear multigrid methods for unstructured mesh solvers. *J. Comp. Phys.*, 175:302–325, 2002.

[192] G. May, F. Iacono, and A. Jameson. A hybrid multilevel method for high-order discretization of the Euler equations on unstructured meshes. *J. Comp. Phys.*, 229:3938–3956, 2010.

[193] P. R. McHugh and D. A. Knoll. Comparison of standard and matrix-free implementations of several Newton-Krylov solvers. *AIAA J.*, 32(12):2394–2400, 1994.

[194] R. C. Mehta. Numerical computation of heat transfer on reentry capsules at Mach 5. *AIAA-Paper 2005-178*, January, 2005.

[195] A. Meister. *Zur zeitgenauen numerischen Simulation reibungsbehafteter, kompressibler, turbulenter Strömungsfelder mit einer impliziten Finite-Volumen-Methode vom Box-Typ*. Dissertation, Technische Hochschule Darmstadt, 1996.

[196] A. Meister. Comparison of different Krylov subspace methods embedded in an implicit finite volume scheme for the computation of viscous and inviscid flow fields on unstructured grids. *J. Comp. Phys.*, 140:311–345, 1998.

[197] A. Meister. *Numerik linearer Gleichungssysteme, Eine Einführung in moderne Verfahren.* Vieweg, Wiesbaden, 1999.

[198] A. Meister and Th. Sonar. Finite-volume schemes for compressible fluid flow. *Surv. Math. Ind.*, 8:1–36, 1998.

[199] A. Meister and C. Vömel. Efficient preconditioning of linear systems arising from the discretization of hyperbolic conservation laws. *Adv. in Comput. Math.*, 14:49–73, 2001.

[200] G. Meurant and J. Duintjer Tebbens. *Krylov Methods for Nonsymmetric Linear Systems.* Springer, Cham, 2020.

[201] C. Michler, E. H. van Brummelen, and R. de Borst. Error-amplification analysis of subiteration-preconditioned GMRES for fluid-structure interaction. *Comp. Meth. Appl. Mech. Eng.*, 195:2124–2148, 2006.

[202] F. Miczek, F. K. Röpke, and P. V. F. Edelmann. New numerical solver for flows at various Mach Numbers. *Astronomy & Astrophysics*, 576(A50), 2015.

[203] M. Minion. Semi-implicit spectral deferred correction methods for ordinary differential equations. *Comm. Math. Sci.*, 1(3):471–500, 2003.

[204] P. Moin and K. Mahesh. DIRECT NUMERICAL SIMULATION: A tool in turbulence research. *Ann. Rev. Fluid Mech.*, 30:539–578, 1998.

[205] A. Monge and P. Birken. On the convergence rate of the Dirichlet–Neumann iteration for unsteady thermal fluid–structure interaction. *Comp. Mech.*, 62(3):525–541, 2018.

[206] A. Monge and P. Birken. A multirate Neumann-Neumann waveform relaxation method for heterogeneous coupled heat equations. *SIAM J. Sci. Comput.*, 41(5):S86–S105, 2019.

[207] A. Monge and P. Birken. Towards a time adaptive Neumann-Neumann waveform relaxation method for thermal fluid-structure interaction. In R. Haynes, S. MacLachlan, X.-C. Cai, L. Halpern, H.H. Kim, A. Klawonn, and O. Widlund, editors, *Domain Decomposition Methods in Science and Engineering XXV.* Springer, 2020.

[208] K. W. Morton and Th. Sonar. Finite volume methods for hyperbolic conservation laws. *Acta Numerica*, pages 155–238, 2007.

[209] N. M. Nachtigal, S. C. Reddy, and L. N. Trefethen. How fast are nonsymmetric matrix iterations? *SIAM J. Matrix Anal. Appl.*, 13(3):778–795, 1992.

[210] C. R. Nastase and D. J. Mavriplis. High-order discontinuous Galerkin methods using an hp-multigrid approach. *J. Comp. Phys.*, 213:330–357, 2006.

[211] C. L. M. H. Navier. Memoire sur les lois du mouvement des fluides. *Mem. Acad. R. Sci. Paris*, 6:389–416, 1823.

[212] J. Nordström, K. Forsberg, C. Adamsson, and P. Eliasson. Finite volume methods, unstructured meshes and strict stability for hyperbolic problems. *Appl. Numer. Math.*, 45(4):453–473, Jun 2003.

[213] J. Nordström and M. Svärd. Well-posed boundary conditions for the Navier-Stokes equations. *SIAM J. Num. Anal.*, 30(3):797, 2005.

[214] M. Ohlberger. A posteriori error estimate for finite volume approximations to singularly perturbed nonlinear convection-diffusion equations. *Numer. Math.*, 87:737–761, 2001.

[215] H. Olsson and G. Söderlind. The approximate Runge-Kutta computational process. *BIT*, 40:351–373, 2000.

[216] S. Osher and S. Chakravarthy. High resolution schemes and the entropy condition. *SIAM J. Num. Anal.*, 21(5):955–984, 1984.

[217] K. Oßwald, A. Siegmund, P. Birken, V. Hannemann, and A. Meister. L2Roe: a low dissipation version of Roe's approximate Riemann solver for low Mach numbers. *Int. J. Num. Meth. Fluids*, 81:71–86, 2016.

[218] B. Owren and M. Zennaro. Order barriers for continuous explicit Runge-Kutta methods. *Math. Comp.*, 56(194):645–661, 1991.

[219] H. Park, R. R. Nourgaliev, R. C. Martineau, and D. A. Knoll. On physics-based preconditioning of the Navier–Stokes equations. *J. Comp. Phys.*, 228(24):9131–9146, 2009.

[220] M. Parsani, M. H. Carpenter, and E. J. Nielsen. Entropy stable wall boundary conditions for the three-dimensional compressible Navier-Stokes equations. *J. Comp. Phys.*, 292:88–113, 2015.

[221] V. C. Patel, W. Rodi, and G. Scheuerer. Turbulence models for near-wall and low-Reynolds number flows: a review. *AIAA Journal*, 23(9):1308–1319, 1985.

[222] O. Peles and E. Turkel. Acceleration methods for multi-physics compressible flow. *J. Comp. Phys.*, 358:201–234, 2018.

[223] J. Peraire and P.-O. Persson. The compact discontinuous Galerkin (CDG) method for elliptic problems. *SIAM J. Sci. Comput.*, 30(4):1806–1824, 2008.

[224] P.-O. Persson and J. Peraire. Newton-GMRES preconditioning for discontinuous Galerkin discretizations of the Navier-Stokes equations. *SIAM J. Sci. Comp.*, 30:2709–2733, 2008.

[225] S. D. Poisson. Memoire sue les equations generales de l'equilibre et du mouvement des corps solides elastiques et des fluides. *J. de l'Ecole Polytechnique de Paris*, 13:139–166, 1831.

[226] N. Qin, D. K. Ludlow, and S. T. Shaw. A matrix-free preconditioned Newton/GMRES method for unsteady Navier-Stokes solutions. *Int. J. Num. Meth. Fluids*, 33:223–248, 2000.

[227] A. Quarteroni and A. Valli. *Domain Decomposition Methods for Partial Differential Equations*. Numerical Mathematics and Scientific Computation. Oxford Science Publications, Oxford, 1999.

[228] K. J. Quint, S. Hartmann, S. Rothe, N. Saba, and K. Steinhoff. Experimental validation of high-order time integration for non-linear heat transfer problems. *Comput. Mech.*, 48(1):81–96, 2011.

[229] J. Rang and L. Angermann. New Rosenbrock W-methods of order 3 for partial differential algebraic equations of index 1. *BIT*, 45:761–787, 2005.

[230] H. Ranocha. Some notes on summation by parts time integration methods. *Results in Applied Mathematics*, 1:100004, 2019.

[231] J. Reisner, V. Mousseau, A. Wyszogrodzki, and D. A. Knoll. A fully implicit hurricane model with physics-based preconditioning. *Monthly Weather Review*, 133:1003–1022, 2005.

[232] J. Reisner, A. Wyszogrodzki, V. Mousseau, and D. Knoll. An efficient physics-based preconditioner for the fully implicit solution of small-scale thermally driven atmospheric flows. *J. Comp. Phys.*, 189(1):30–44, 2003.

[233] F. Reitsma, G. Strydom, J. B. M. de Haas, K. Ivanov, B. Tyobeka, R. Mphahlele, T. J. Downar, V. Seker, H. D. Gougar, D. F. Da Cruz, and U. E. Sikik. The PBMR steadystate and coupled kinetics core thermal-hydraulics benchmark test problems. *Nuclear Engineering and Design*, 236(5-6):657–668, 2006.

[234] W. C. Rheinboldt. *Methods for Solving Systems of Nonlinear Equations*. SIAM, 2nd edition, 1998.

[235] F. Rieper. A low-Mach number fix for Roe's approximate Riemann solver. *J. Comp. Phys.*, 230:5263–5287, 2011.

[236] F. Rieper and G. Bader. Influence of cell geometry on the accuracy of upwind schemes in the low Mach number regime. *J. Comp. Phys.*, 228:2918–2933, 2009.

[237] C.-C. Rossow. Convergence acceleration for solving the compressible Navier–Stokes equations. *AIAA J.*, 44:345–352, 2006.

[238] C.-C. Rossow. Efficient computation of compressible and incompressible flows. *J. Comp. Phys.*, 220(2):879–899, 2007.

[239] B. Rüth, B. Uekermann, M. Mehl, P. Birken, A. Monge, and H-J. Bungartz. Quasi-Newton Waveform Iteration for Partitioned surface-coupled multiphysics Applications. *I. J. Num. Meth. Eng.*, Online fir, 2020.

[240] Y. Saad. A flexible inner-outer preconditioned GMRES algorithm. *SIAM J. Sci. Comput.*, 14(2):461–469, 1993.

[241] Y. Saad. *Iterative Methods for Sparse Linear Systems*. PWS Publishing Company, Boston, 1996.

[242] Y. Saad and M. H. Schultz. GMRES: A generalized minimal residual algorithm for solving nonsymmetric linear systems. *SIAM J. Sci. Stat. Comput.*, 7:856–869, 1986.

[243] M. Sala and R. S. Tuminaro. A new Petrov-Galerkin smoothed aggregation preconditioner for nonsymmetric linear systems. *SIAM J. Sci. Comput.*, 31(1):143–166, 2008.

[244] H. Schlichting and K. Gersten. *Boundary Layer Theory*. Springer, 8th revise edition, 2000.

[245] S. Schüttenberg, M. Hunkel, U. Fritsching, and H.-W. Zoch. Controlling of distortion by means of quenching in adapted jet fields. *Materialwissenschaft und Werkstofftechnik*, 37(1):92–96, 2006.

[246] L. F. Shampine. *Numerical Solution of Ordinary Differential Equations*. Springer, 1994.

[247] C.-W. Shu. Total-variation diminishing time discretizations. *SIAM J. Sci. Stat. Comp.*, 9:1073–1084, 1988.

[248] C.-W. Shu and S. Osher. Efficient implementation of essentially nonoscillatory shock-capturing schemes. *J. Comp. Phys.*, 77:439–471, 1988.

[249] V. Simoncini and D. B. Szyld. Recent computational developments in Krylov subspace methods for linear systems. *Num. Lin. Algebra with Appl.*, 14:1–59, 2007.

[250] J. Smagorinsky. General circulation experiments with the primitive equations. *Monthly Weather Review*, 91:99–165, 1963.

[251] G. Söderlind. Digital filters in adaptive time-stepping. *ACM TOMS*, 29(1):1–26, 2003.

[252] G. Söderlind. The logarithmic norm. History and modern theory. *BIT Numerical Mathematics*, 46(3):631–652, 2006.

[253] G. Söderlind. Time-step selection algorithms: Adaptivity, control, and signal processing. *Appl. Num. Math.*, 56:488–502, 2006.

[254] G. Söderlind and L. Wang. Adaptive time-stepping and computational stability. *J. Comp. Appl. Math.*, 185:225–243, 2006.

[255] G. Söderlind and L. Wang. Evaluating numerical ODE/DAE methods, algorithms and software. *J. Comp. Appl. Math.*, 185:244–260, 2006.

[256] P. Sonneveld and M. B. van Gijzen. IDR(s): A family of simple and fast algorithms for solving large nonsymmetric systems of linear equations. *SIAM J. Sci. Comp.*, 31(2):1035–1062, 2008.

[257] P. R. Spalart and S. R. Allmaras. A one-equation turbulence model for aerodynamic flows. *AIAA 30th Aerospace Science Meeting*, 92–0439, 1992.

[258] P. R. Spalart, W. Jou, M. Strelets, and S. R. Allmaras. Comments on the feasibility of LES for wings, and on a hybrid RANS/LES approach. In *1st ASOSR CONFERENCE on DNS/LES. Arlington, TX*, 1997.

[259] S. P. Spekreijse. Multigrid solution of monotone second-order discretizations of hyperbolic conservation laws. *Math. Comp.*, 49(179):135–155, 1987.

[260] A. St-Cyr and D. Neckels. A fully implicit Jacobian-free high-order discontinuous Galerkin mesoscale flow solver. In *Proceedings of the 9th International Conference on Computational Science*, pages 243–252, Berlin, Heidelberg, 2009. Springer.

[261] G. G. Stokes. On the theories of the internal friction of fluids in motion. *Trans. Camb. Phil. Soc.*, 8:287–305, 1845.

[262] G. Strang. On the construction and comparison of difference schemes. *SIAM J. Num. Anal.*, 5(3):506–517, 1968.

[263] P. Stratton, I. Shedletsky, and M. Lee. Gas quenching with helium. *Solid State Phenomena*, 118:221–226, 2006.

[264] K. Strehmel and R. Weiner. *Linear-Implizite Runge-Kutta-Methoden und ihre Anwendung*. Teubner, Stuttgart, 1992.

[265] M. Svärd, M. H. Carpenter, and J. Nordström. A stable high-order finite difference scheme for the compressible Navier-Stokes equations, far-field boundary conditions. *J. Comp. Phys.*, 225(1):1020–1038, 2007.

[266] M. Svärd and J. Nordström. Review of summation-by-parts schemes for initial-boundary-value problems. *J. Comp. Phys.*, 268:17–38, 2014.

[267] M. Svärd and H. Özcan. Entropy-stable schemes for the Euler equations with far-field and wall boundary conditions. *J. Sci. Comput.*, 58(1):61–89, 2014.

[268] R. C. Swanson and E. Turkel. Analysis of a fast iterative method in a dual time algorithm for the Navier-Stokes equations. In J. Eberhardsteiner, editor, *European Congress on Computational Methods and Applied Sciences and Engineering (ECCOMAS 2012)*, 2012.

[269] R. C. Swanson, E. Turkel, and C.-C. Rossow. Convergence acceleration of Runge–Kutta schemes for solving the Navier–Stokes equations. *J. Comp. Phys.*, 224(1):365–388, 2007.

[270] R. C. Swanson, E. Turkel, and S. Yaniv. Analysis of a RK/implicit smoother for multigrid. In A. Kuzmin, editor, *Computational Fluid Dynamics 2010*. Springer, Berlin, Heidelberg, 2011.

[271] T. Tang and Z.-H. Teng. Error bounds for fractional step methods for conservation laws with source terms. *SIAM J. Num. Anal.*, 32(1):110–127, 1995.

[272] Open MPI Development team. OpenMPI.

[273] K. Thompson. Time dependent boundary conditions for hyperbolic systems. *J. Comp. Phys.*, 68:1–24, 1987.

[274] B. Thornber, A. Mosedale, D. Drikakis, and D. Youngs. An improved reconstruction method for compressible flows with low Mach number features. *J. Comp. Phys.*, 227:4873–4894, 2008.

[275] V. A. Titarev and E. F. Toro. ADER: Arbitrary high order Godunov approach. *J. Sci. Comp.*, 17(1-4):609–618, 2002.

[276] M. Tokman. A new class of exponential propagation iterative methods of Runge–Kutta type (EPIRK). *J. Comp. Phys.*, 230(24):8762–8778, 2011.

[277] E. F. Toro. *Riemann Solvers and Numerical Methods for Fluid Dynamics*. Springer, Berlin, Heidelberg, New York, 2. edition, 1999.

[278] A. Toselli and O. B. Widlund. *Domain Decomposition Methods— Algorithms and Theory*, volume 34. Springer, Berlin, Heidelberg, 2004.

[279] L. N. Trefethen. Pseudospectra of linear operators. *SIAM Review*, 39 (3):383–406, 1997.

[280] D. Tromeur-Dervout and Y. Vassilevski. Choice of initial guess in iterative solution of series of systems arising in fluid flow simulations. *J. Comp. Phys.*, 219:210–227, 2006.

[281] U. Trottenberg, C. W. Oosterlee, and S. Schüller. *Multigrid*. Elsevier Academic Press, 2001.

[282] S. Turek. *Efficient Solvers for Incompressible Flow Problems: An Algorithmic and Computational Approach*. Springer, Berlin, 1999.

[283] H. A. van der Vorst. BI-CGSTAB: A fast and smoothly converging variant of BI-CG for the solution of nonsymmetric linear systems. *SIAM J. Sci. Stat. Comput.*, 13:631–644, 1992.

[284] H. A. van der Vorst. *Iterative Krylov Methods for Large Linear Systems*, volume 13 of *Cambridge Monographs on Applied and Computational Mathematics*. Cambridge University Press, Cambridge, 2003.

[285] H. A. van der Vorst and C. Vuik. GMRESR: a family of nested GMRES methods. *Num. Lin. Algebra with Appl.*, 1(4):369–386, 1994.

[286] B. van Leer. Flux-vector splitting for the Euler equations. In E. Krause, editor, *Eighth International Conference on Numerical Methods in Fluid Dynamics*, number 170 in Lecture Notes in Physics, pages 507–512, Berlin, 1982. Springer Verlag.

[287] B. van Leer, C.-H. Tai, and K. G. Powell. Design of optimally smoothing multi-stage schemes for the Euler equations. In *AIAA 89-1933-CP*, pages 40–59, 1989.

[288] A. van Zuijlen and H. Bijl. Implicit and explicit higher order time integration schemes for structural dynamics and fluid-structure interaction computations. *Comp. Struct.*, 83:93–105, 2005.

[289] A. van Zuijlen, A. de Boer, and H. Bijl. Higher-order time integration through smooth mesh deformation for 3D fluid–structure interaction simulations. *J. Comp. Phys*, 224:414–430, 2007.

[290] R. S. Varga. *Matrix Iterative Analysis*, volume 27 of *Series in Computational Mathematics*. Springer, New York, Berlin, Heidelberg, 2000.

[291] V. Venkatakrishnan. Convergence to steady state solutions of the Euler equations on unstructured grids with limiters. *J. Comp. Phys.*, 118:120–130, 1995.

[292] J. Vierendeels, L. Lanoye, J. Degroote, and P. Verdonck. Implicit coupling of partitioned fluid-structure interaction problems with reduced order models. *Comp. Struct.*, 85:970–976, 2007.

[293] P. Vijayan and Y. Kalinderis. A 3D finite-volume scheme for the Euler equations on adaptive tetrahedral Grids. *J. Comp. Phys.*, 113:249–267, 1994.

[294] P. E. Vincent, P. Castonguay, and A. Jameson. A new class of high-order energy stable flux reconstruction schemes. *J. Sci. Comp.*, 47:50–72, 2011.

[295] P. E. Vincent and A. Jameson. Facilitating the adoption of unstructured high-order methods amongst a wider community of fluid dynamicists. *Math. Model. Nat. Phenom.*, 6(3):97–140, 2011.

[296] Y. Wada and M.-S. Liou. A flux splitting scheme with high-resolution and robustness for discontinuities. *AIAA Paper 94-0083*, 1994.

[297] K. Wang, W. Xue, H. Lin, S. Xu, and W. Zheng. Updating preconditioner for iterative method in time domain simulation of power systems. *Science China Technological Sciences*, 54(4):1024–1034, feb 2011.

[298] L. Wang and D. J. Mavriplis. Implicit solution of the unsteady Euler equations for high-order accurate discontinuous Galerkin discretizations. *J. Comp. Phys.*, 225:1994–2015, 2007.

[299] U. Weidig, N. Saba, and K. Steinhoff. Massivumformprodukte mit funktional gradierten Eigenschaften durch eine differenzielle thermo-mechanische Prozessführung. *WT-Online*, pages 745–752, 2007.

[300] R. Weiner, B. A. Schmitt, and H. Podhaisky. ROWMAP a ROW-code with Krylov techniques for large stiff ODEs. *Appl. Num. Math.*, 25:303–319, 1997.

[301] P. Wesseling. *Principles of Computational Fluid Dynamics*. Springer, 2001.

[302] P. Wesseling. *An Introduction to Multigrid Methods*. R T Edwards Inc, 2004.

[303] L. B. Wigton, N. J. Yu, and D. P. Young. GMRES acceleration of computational fluid dynamics codes. *AIAA Paper 85-1494*, 1985.

[304] F. Zahle, N. N. Soerensen, and J. Johansen. Wind turbine rotor-tower interaction using an incompressible overset grid method. *Wind Energy*, 12:594–619, 2009.

[305] Q. Zhang and C.-W. Shu. Error estimates to smooth solutions of Runge-Kutta discontinuous Galerkin method for symmetrizable systems of conservation laws. *SIAM J. Num. Anal.*, 44(4):1703–1720, 2006.

Index

Residual averaging, 115
Restriction, 105, 107, 116
Reynolds number, 14
Reynolds' transport theorem, 10
Richardson iteration, 102
Rosenbrock method, 82, 166
Rotational invariance, 33
Runge-Kutta method, 75

SBP-SAT, 59
SDIRK method, 79, 180
Second law of thermodynamics, 19
SGS method, 102, 115
Simplified Newton's method, 129
Slope limiter, 47
Smoothed aggregation, 115
Smoother, 107, 110
Smoothing factor, 118
Source terms, 16, 45
Space-time method, 91
Specific gas constant, 13
Specific heat coefficient, 13
Spectral radius, 121
Speed of sound, 14
Splitting schemes, 89
SSOR method, 103
SSP method, 68
Stability, 59, 66
Stability function, 67, 76, 77, 79, 111
Stability region, 67
Stationary linear method, 101, 111
Step size controller, 87
Stiff problem, 72
Stiffly accurate, 76
Stokes hypothesis, 12
Strang splitting, 90
Stress tensor, 12
Strong stability, 68
Subsonic flow, 18
Supersonic flow, 18
Sutherland law, 13

Temperature, 12, 13
Termination criterion, 100, 102, 131, 164, 181

Theorem of Lax-Wendroff, 43
Thermal conductivity, 12
Time adaptivity, 86, 180
Tolerance, 87, 130, 164, 166
Tolerance, iteration, 100
Tolerance, time adaptivity, 87
Total variation, 43
Transonic flow, 18
Trapezoidal rule, 77
Turbulent flow, 22
TV stability, 45
TVD scheme, 45
TVD stability, 68

V-cycle, 108
Velocity, 10
Venkatakrishnan limiter, 48
Viscosity, dynamic, 12, 13
Von Neumann stability analysis, 70

W method, 82
W-cycle, 108
Weak entropy solution, 19
Weak solution, 19